SpringerBriefs in Applied Sciences and Technology

SpringerBriefs present concise summaries of cutting-edge research and practical applications across a wide spectrum of fields. Featuring compact volumes of 50 to 125 pages, the series covers a range of content from professional to academic.

Typical publications can be:

- A timely report of state-of-the art methods
- An introduction to or a manual for the application of mathematical or computer techniques
- A bridge between new research results, as published in journal articles
- A snapshot of a hot or emerging topic
- An in-depth case study
- A presentation of core concepts that students must understand in order to make independent contributions

SpringerBriefs are characterized by fast, global electronic dissemination, standard publishing contracts, standardized manuscript preparation and formatting guidelines, and expedited production schedules.

On the one hand, **SpringerBriefs in Applied Sciences and Technology** are devoted to the publication of fundamentals and applications within the different classical engineering disciplines as well as in interdisciplinary fields that recently emerged between these areas. On the other hand, as the boundary separating fundamental research and applied technology is more and more dissolving, this series is particularly open to trans-disciplinary topics between fundamental science and engineering.

Indexed by EI-Compendex, SCOPUS and Springerlink.

Tiziana Ferrante · Marco Sacco
Editors

Habitable Future

Smart Spaces, Objects, and Devices to
Support Aging

Editors
Tiziana Ferrante
Department of Planning,
Design, Architectural Technology
Sapienza University of Rome
Rome, Italy

Marco Sacco
Institute of Intelligent Industrial
Technologies and Systems for Advanced
Manufacturing (STIIMA)
National Research Council of Italy (CNR)
Lecco, Italy

ISSN 2191-530X   ISSN 2191-5318 (electronic)
SpringerBriefs in Applied Sciences and Technology
ISBN 978-3-031-95734-5   ISBN 978-3-031-95735-2 (eBook)
https://doi.org/10.1007/978-3-031-95735-2

This work was supported by Consiglio Nazionale delle Ricerche.

© The Editor(s) (if applicable) and The Author(s) 2025. This book is an open access publication.

Open Access This book is licensed under the terms of the Creative Commons Attribution 4.0 International License (http://creativecommons.org/licenses/by/4.0/), which permits use, sharing, adaptation, distribution and reproduction in any medium or format, as long as you give appropriate credit to the original author(s) and the source, provide a link to the Creative Commons license and indicate if changes were made.
The images or other third party material in this book are included in the book's Creative Commons license, unless indicated otherwise in a credit line to the material. If material is not included in the book's Creative Commons license and your intended use is not permitted by statutory regulation or exceeds the permitted use, you will need to obtain permission directly from the copyright holder.
The use of general descriptive names, registered names, trademarks, service marks, etc. in this publication does not imply, even in the absence of a specific statement, that such names are exempt from the relevant protective laws and regulations and therefore free for general use.
The publisher, the authors and the editors are safe to assume that the advice and information in this book are believed to be true and accurate at the date of publication. Neither the publisher nor the authors or the editors give a warranty, expressed or implied, with respect to the material contained herein or for any errors or omissions that may have been made. The publisher remains neutral with regard to jurisdictional claims in published maps and institutional affiliations.

This Springer imprint is published by the registered company Springer Nature Switzerland AG
The registered company address is: Gewerbestrasse 11, 6330 Cham, Switzerland

If disposing of this product, please recycle the paper.

Preface

Real progress happens only when advantages of a new technology become available to everybody.

So says an aphorism by Henry Ford.

This statement gains particular depth when considered in light of one of the most radical transformations of our time: population ageing. If innovation is to be a true driver of equity and social development, we must ask not only whom it serves, but also what effects it produces.

In this context, emerging technologies–from smart assistive devices to artificial intelligence–can and must contribute to the design of living environments that support the dynamic process of ageing: flexible, safe, and adaptable spaces, conceived to empower and never to constrain.

Innovation becomes a genuine progress when it is translated into accessibility, care, and improved quality of living.

We live in an era where longevity is no longer an exception but a widespread global condition. The steady rise in life expectancy is a structural social phenomenon of our present, with profound economic and healthcare implications, one that is poised to profoundly reshape the future of our societies and compels us to rethink how we inhabit time, space, and human relationships.

It is within this context that the present volume takes shape. Rather than approaching ageing as a challenge to be "contained", it frames it as an opportunity to reimagine our collective future. The contributions collected here offer a broad, multi-faceted, and multidisciplinary reflection on the implications of longevity, giving voice to an epochal shift that too often remains in the background of public debate–despite already constituting a pressing issue.

Population ageing is no longer solely a demographic concern: it is a cultural, social, economic, and design challenge. Addressing it means envisioning new ways of living, new services, new spatial models, as well as new forms of social connection and cohesion.

This volume is conceived as a response to that challenge, because today, more than ever, we need to develop visions and tools capable of supporting this ongoing transformation.

The contents collected in this volume stem from a research pathway rooted in a national project that is both ambitious and innovative. They represent the first outcomes of the Age-It program, funded through the Italian National Recovery and Resilience Plan (PNRR), and in particular of Work Package 1 within Spoke 9 "Advanced Gerontechnologies for Active and Healthy Ageing". This project places the transformation of ageing into an active and dignified phase of life at its core, to be supported through bold visions and innovative tools.

The volume was conceived as a platform for connecting diverse forms of knowledge and disciplinary perspectives, aiming to offer concrete, thoughtful, empathetic, and rigorous responses to the evolving needs of an ageing population. At the same time, it seeks to guide readers, scholars, professionals, and public decision-makers toward a renewed awareness: that ageing, when accompanied by fair and inclusive technological innovation–particularly directed at the most vulnerable individuals and disadvantaged contexts–can represent an opportunity rather than a problem, ensuring that no one is left behind.

What particularly engages the reader is the way the issue of ageing and living environments is addressed. How can we design inclusive, safe, and flexible spaces that dynamically support the various phases of later life? How can we develop technologies that make a tangible difference in the everyday lives and care needs of older adults, without intruding on their privacy or replacing human relationships? This volume seeks to answer such questions by promoting the creation of physical and social environments in which older adults are not merely "contained," but actively supported in living, acting, and making choices.

Expanding on these reflections, the thirteen chapters explore–through diverse yet complementary perspectives–socio-economic considerations, technological solutions, and project-oriented design and architectural approaches.

In particular, this volume introduces a new vision for longevity-focused technologies, capable of responding to the real needs of older adults, informed by behavioral science approaches, such as those underpinning the Internet of Caring Things ("From Internet of Things to Internet of Caring Things: A Paradigm Shift for Healthy Longevity").

In line with this perspective, a specific focus on advancements in life sciences and biotechnology underscores the importance of cross-disciplinary collaboration, structured within multidisciplinary hubs, as essential for addressing current demographic shifts ("Technology for Life").

The opportunities of the silver economy, understood as an innovative ecosystem designed to meet the needs of the aging population, are explored as a strategic driver for inclusive growth through the integration of technological innovation, public policy, and private initiative, especially in the realm of digital health ("Silver Economy for Innovation and Inclusivity: Sustainable Environments for Life Independency").

This orientation is further reflected in the use of human-centered design methodologies, aimed at capturing users' needs and expectations as a foundation for future design strategies for innovative and accessible gerontechnologies ("The Needs of the Over-55 Population and Design Strategies for the Development of Gerontechnologies for Active and Healthy Ageing").

Equally essential is the direct involvement of users in the design and testing phases, to better understand the complex relationship between aging individuals and technology, and to identify effective strategies for improving both acceptance and adoption ("Technologies for Elderly Care: Relevance and Challenges of User Engagement in the Design and Test Phases").

The dynamics of innovation in the health sector, particularly in relation to aging, are further examined through patent data analysis, which highlights the growing prevalence of digital technologies and their increasingly interdisciplinary nature ("New Technologies for an Aging Population: Trends and Opportunities").

Moreover, field testing of systems that combine sensing technologies and data management provides practical insights into the implementation of gerontechnologies ("Autonomous Devices for Elderly Monitoring and Assistance: Enhancing Quality of Life with Custom Sensors and Yottabyte-Scale Solutions").

Complementing these efforts, the development of new decision-support tools based on ontological frameworks offers a means to tailor living environments more effectively, taking into account individual health conditions, spatial constraints, and the technical features of available technologies ("Reconfiguring and Customizing Living Environments with Knowledge: the *Age-It* Decision Support System").

The design of spaces that integrate new technologies also acquires a relational dimension, aimed at reinforcing community ties, particularly within emerging models of territorial welfare ("Social Capital-Rich Places for Healthy Aging")—as well as upholding the right to *aging in place* through housing models conceived as integrated services, potentially extendable to collective living arrangements ("The Right to Grow Old in Your Own Home").

Indeed, the volume explores three key contexts for *aging-in-place*: Housing, Community Health Centers, and Social Day Care Centers.

With regard to Community Health Centers, the volume presents design guidelines for waiting areas that integrate technologies and sensory environments to support healthy and active aging, as illustrated through the experience of a pilot project ("Waiting Spaces in Primary Care for Healthy Ageing: Applying Design Guidelines to the Project").

In light of the pressing need to improve the integration of assistive technologies within the home, particular attention is dedicated to the adaptability of domestic environments in the context of home care. This includes defining both spatial and technological features, as well as providing operational design guidelines for professionals ("Adaptability of Housing for Home Care: Strategies and Design Solutions for Assistive Spaces and Technologies").

Finally, for Social Day Care Centers, the volume proposes architectural strategies inspired by successful European models, with the aim of contributing to the development of a more inclusive and effective proximity-based welfare system ("Social Day Care Centre: A New Architectural Model to Improve Elderly's Quality of Life").

In conclusion, this volume does not merely aim to describe an ongoing transformation, but to actively contribute to designing it. The future is not something that simply happens: it is a shared challenge that concerns us all, and it can only become reality through the commitment and collaboration of multidisciplinary expertise.

Good Reading

Rome, Italy Tiziana Ferrante
Lecco, Italy Marco Sacco

Competing Interests

This volume was developed within the project funded by Next Generation EU—"Age-It—Ageing well in an ageing society" project (PE0000015), National Recovery and Resilience Plan (NRRP)—PE8—Mission 4, C2, Intervention 1.3".

Contents

From Internet of Things to Internet of Caring Things: A Paradigm Shift for Healthy Longevity 1
Nic Palmarini and Jennine Jonczyk

Technology for Life 7
Massimiliano Boggetti

Silver Economy for Innovation and Inclusivity: Sustainable Environments for Life Independency 11
Felice Lopane

The Needs of the Over-55 Population and Design Strategies for the Development of Gerontechnologies for Active and Healthy Ageing 17
Mattia Pistolesi and Francesca Tosi

Technologies for Elderly Care: Relevance and Challenges of User Engagement in the Design and Test Phases 33
Lorena Rossi, Valentina Tombolesi, Rachele Alessandra Marziali, Eleonora Bonifazi, and Vera Stara

New Technologies for an Aging Population: Trends and Opportunities 47
Matteo Romagnoli and Maria Luisa Mancusi

Autonomous Devices for Elderly Monitoring and Assistance: Enhancing Quality of Life with Custom Sensors and Yottabyte-Scale Solutions 61
Livio D'Alvia, Christian Napoli, and Zaccaria Del Prete

Reconfiguring and Customizing Living Environments with Knowledge: The *Age-It* Decision Support System 73
Daniele Spoladore, Atieh Mahroo, and Marco Sacco

Social Capital-Rich Places for Healthy Aging 89
Roberto Di Monaco, Silvia Pilutti, and Marzia Ravazzini

The Right to Grow Old in Your Own Home 95
Claudio Falasca

Waiting Spaces in Primary Care for Healthy Ageing: Applying Design Guide-lines to the Project 101
Elena Bellini and Nicoletta Setola

Adaptability of Housing for Home Care: Strategies and Design Solutions for Assistive Spaces and Technologies 119
Federica Romagnoli, Teresa Villani, and Tiziana Ferrante

Social Day Care Centre: A New Architectural Model to Improve Elderly's Quality of Life 141
Maria Argenti, Fabio Cutroni, Domizia Mandolesi,
Anna Bruna Menghini, Maura Percoco, and Francesca Sarno

From Internet of Things to Internet of Caring Things: A Paradigm Shift for Healthy Longevity

Nic Palmarini and Jennine Jonczyk

Abstract The global ageing population presents unique challenges and opportunities for technological innovation. This paper introduces the Internet of Caring Things (IoCT) as a novel evolution of the Internet of Things (IoT), reframing the approach to technology development for healthy ageing and longevity. Unlike traditional IoT applications or age-specific technologies, IoCT focuses on what people genuinely care about in their daily lives, integrating data-driven insights to support well-being across the lifespan. Through analysis of behavioral factors, market trends, and real-world case studies, this paper demonstrates how IoCT represents a significant paradigm shift in addressing healthy longevity challenges, moving from "caring about people" to "what people care about".

Keywords IoT · IoCT · Ageing · Longevity · Technology · Behaviour · Devices · Wearable · Health · Healthspan · Agetech · Gerontech

1 Introduction: The Longevity Paradigm

By 2050, the global population will include approximately 2.1 billion people over the age of 60, representing 21.3% of the total population [1]. This demographic shift constitutes one of the most significant megatrends impacting commercial enterprises and social interactions worldwide. Despite the availability of technologies designed to support older adults, adoption rates remain surprisingly low.

Less than one in three people over 50 currently use health apps, and only 28% of people with diabetes use technology to track blood sugar levels [2].

This adoption gap reveals a fundamental disconnect in current approaches to technology development for ageing populations. As veteran technology leaders observe:

N. Palmarini (✉) · J. Jonczyk
UK National Innovation Centre for Ageing, Newcastle University, Newcastle Upon Tyne, UK
e-mail: nic.palmarini@newcastle.ac.uk

J. Jonczyk
e-mail: jennine.jonczyk@newcastle.ac.uk

© The Author(s) 2025
T. Ferrante and M. Sacco (eds.), *Habitable Future*,
SpringerBriefs in Applied Sciences and Technology,
https://doi.org/10.1007/978-3-031-95735-2_1

"Technology is relatively easy. It's people that are hard" [3]. The challenge extends beyond creating functional technology—it requires understanding human factors that determine success or failure, particularly for older adults who may rely on existing mental models rather than adapting to new information systems.

This article introduces the Internet of Caring Things (IoCT) as a paradigm shift that addresses these challenges by focusing not on age-specific needs but on what people genuinely care about throughout their lives.

2 From IoT to IoCT: A Conceptual Framework

2.1 Limitation of Current Approaches

Current technological approaches to ageing populations are often constrained by three key limitations:

1. **Focus on disability rather than capability**: Most "AgeTech" solutions emphasize solving problems related to age-related decline rather than enhancing existing capabilities or preventing issues before they arise.
2. **Lack of perceived benefit**: Many older adults do not see sufficient value in technology adoption to overcome learning barriers, particularly when solutions are imposed rather than integrated into existing habits.
3. **Siloed development**: Technologies are often developed for isolated problems without considering how they integrate into holistic living experiences.

The Internet of Caring Things addresses these limitations by fundamentally reorienting how we conceptualize technology for longevity.

2.2 Defining the Internet of Caring Things

The Internet of Caring Things is defined as "a network of connected devices and services that actively care for people and what matters to them" [4]. This framework encompasses three critical dimensions:

1. Technologies that reflect what people genuinely care about in their daily lives (pets, hobbies, family, home, etc.).
2. Connected systems that facilitate meaningful relationships and social engagement.
3. Data-driven solutions that add value to healthy ageing through prevention, support, and care.

The significant shift in IoCT lies in its approach to addressing behavioral limiting factors by leveraging existing interests rather than creating new behaviors. Instead of asking, "Why should we care about things that should care about us if we don't care about them?", IoCT embeds supportive technology within contexts, processes, and activities people already value.

3 Behavioral Foundations of IoCT

The IoCT approach is rooted in behavioral science principles that acknowledge common barriers to technology adoption:

- **Fear of the unknown**: IoCT introduces technologies through familiar contexts and stakeholders.
- **Difficulty focusing**: Rather than requiring specific focus, IoCT leverages and maximizes existing attention patterns.
- **Power of habit**: IoCT maps existing habits and delivers support through established routines.
- **Time constraints**: Solutions are integrated into daily flow without adding new activities.
- **Rushing**: The pace of interaction matches existing life rhythms.
- **Procrastination**: By eliminating new activities, IoCT reduces barriers to engagement.
- **Goal setting**: Goals are derived from what people already care about.

This behavioral approach recognizes that "when there's something (older) adults want to do, nothing is going to get in the way of an older adult using technology" [5]. By embedding technology within existing interests—such as gaming ($545.98 billion market), pets ($232 billion market), or DIY home improvement ($1,278.0 billion market)—IoCT harnesses intrinsic motivation rather than imposing external requirements.

4 IoCT Implementation: Case Studies

The practical application of IoCT principles can be observed across various domains, demonstrating its versatility and effectiveness:

Informetis: Behavioral Insights Through Power Consumption. Informetis developed a non-intrusive system that monitors household power consumption using AI algorithms to deduce which appliances are being used and for how long. Rather than requiring users to adopt new monitoring behaviors, the system maps routine behaviors and raises alerts when deviations occur.

In a twelve-week implementation with 17 participants, the system provided valuable insights for ageing in place while preserving privacy and dignity. This approach demonstrated how passive monitoring through everyday activities could provide behavioral mapping and risk assessment without disrupting established routines [6].

Perro: Pet and Owner Well-being Connection. Recognizing the $232 billion pet industry and the emotional connection between owners and pets, Perro developed a connected digital health platform that monitors both pet and owner well-being. The system leverages the natural care relationship between pets and humans, encouraging physical activity through dog walking while monitoring health indicators for both. This approach demonstrates how caring for something else (pets) can motivate self-care behaviors, creating a virtuous cycle of increased activity and well-being [7].

Circadacare: Lighting for Healthy Living. Circadacare developed a novel lighting system that supports healthy circadian rhythms through natural lighting patterns. Rather than requiring users to actively engage with health monitoring, the solution is embedded in the home environment, passively supporting sleep cycles and overall well-being. This case illustrates how environmental modifications can support health outcomes without requiring behavioral change, earning significant funding ($2M) and recognition for its alignment with natural human patterns [8].

5 Future Directions and Implications

The Internet of Caring Things reflects a shift in our approach to the development of technology for the betterment of longevity. This paradigm change introduces a few critical implications that have profound consequences for research, business development, policy-making, and civic engagement.

Market Integration and Economic Impact. IoCT creates business opportunities by addressing real needs through natural behaviors rather than imposing solutions. The approach has demonstrated commercial viability with over 105 SMEs empowered, four new products launched, and £10M+ in funds raised within two years of implementation [8].

Community Engagement Models. New models of citizen engagement have emerged through IoCT initiatives, including Voice Italia community, market-based interactions, and community toolboxes that allow citizens to explore smart sensor technologies in daily life. These engagement models bridge the gap between technology developers and users, creating sustainable ecosystems of care.

From Smart Homes to Longevity Homes. The concept of "smart homes" is evolving toward "longevity homes" that integrate supportive technologies into living environments without stigmatizing ageing. Projects like Bernicia demonstrate how housing providers can implement IoCT principles to create living spaces that support healthy ageing through embedded, non-intrusive technologies.

6 Conclusion

The Internet of Caring Things represents a significant paradigm shift in addressing the challenges of global population ageing. By focusing on what people genuinely care about rather than imposing technology solutions, IoCT offers a more human-centered approach to healthy longevity. The integration of data-driven insights with everyday objects and activities creates opportunities for prevention, support, and care without requiring new behaviors or stigmatizing ageing.

As Antoine de Saint-Exupéry noted, "What is essential is invisible to the eye." The IoCT approach recognizes that the most effective technologies may be those that blend seamlessly into the fabric of daily life, supporting well-being through the things people already value and enjoy. This shift from technology-centric to human-centric design represents a promising direction for creating more inclusive, effective solutions for healthy longevity in an ageing world.

References

1. United Nations. (2019). *World population prospects 2019: Highlights.* https://population.un.org/wpp/assets/Files/WPP2019_Highlights.pdf
2. Institute for Healthcare Policy and Innovation. (2023). *Health apps could help older adults with anything from sleep to diabetes, but most don't use them.* https://ihpi.umich.edu/news-events/news/health-apps-could-help-older-adults-anything-sleep-diabetes-most-dont-use-them
3. Deloitte. (2022). *The human side of tech: Driving behavioral change.* https://deloitte.wsj.com/cio/the-human-side-of-tech-driving-behavioral-change-1522900929
4. Ncleus. (2022). *Breaking ageing stigmas with the internet of caring things.* https://ncleus.com/2022/04/breaking-ageing-stigmas-with-the-internet-of-caring-things/Pew; Research Center. *Technology use among older adults.* https://www.pewresearch.org/short-reads/2022/01/13/share-of-those-65-and-older-who-are-tech-users-has-grown-in-the-past-decade/
5. Newcastle University. (2022, 29 November). *NICA working with Informetis—Business and partnerships.* Newcastle University. Retrieved April 15, 2025, from https://www.ncl.ac.uk/business-and-partnerships/nica-working-with-informetis
6. American Academy of Neurology. (2022, 23 February). *Do pets have a positive effect on your brain health?* ScienceDaily. Retrieved April 15, 2025, from https://www.sciencedaily.com/releases/2022/02/220223210035.htm; UK National Innovation Centre for Ageing. (2022). *My dog case study: Your pet's health, therefore your health.*
7. Circadacare. (2025). *Solutions.* Circadacare. Retrieved April 15, 2025, from https://www.circadacare.com/
8. UK National Innovation Centre for Ageing. (2023). *Main outcomes after 2 years (so far).* https://ioct.uknica.co.uk/

Open Access This chapter is licensed under the terms of the Creative Commons Attribution 4.0 International License (http://creativecommons.org/licenses/by/4.0/), which permits use, sharing, adaptation, distribution and reproduction in any medium or format, as long as you give appropriate credit to the original author(s) and the source, provide a link to the Creative Commons license and indicate if changes were made.

The images or other third party material in this chapter are included in the chapter's Creative Commons license, unless indicated otherwise in a credit line to the material. If material is not included in the chapter's Creative Commons license and your intended use is not permitted by statutory regulation or exceeds the permitted use, you will need to obtain permission directly from the copyright holder.

Technology for Life

Massimiliano Boggetti

Abstract This paper examines rapid technological advancements in life sciences, highlighting their transformative effects on healthcare and patient outcomes. Emphasizing interdisciplinary collaboration and significant investments in research, the study discusses innovative solutions such as AI-driven diagnostics, genetic engineering, and human-centric medical devices. Addressing demographic shifts, global healthcare disparities, and leveraging collaborative hubs like Italy's ALISEI cluster, it argues that continuous innovation not only improves global health equity and economic sustainability but also opens pathways toward a potential evolution of human capabilities.

Keywords Life sciences · Technological innovation · Healthcare · Interdisciplinary collaboration · Research investment · Human-centric design

1 Introduction

We are currently experiencing an extraordinary period characterized by rapid technological innovation. One particularly critical area of advancement is in the life sciences, which encompasses over 22,000 pharmaceutical products and more than 1.5 million medical devices.

These innovations are vital for the well-being of patients worldwide on a daily basis, facilitating early diagnosis, personalized medicine, and improved treatment outcomes. The intersection of technology and healthcare has led to significant improvements in patient care and quality of life.

M. Boggetti (✉)
ALISEI Advanced Life Sciences in Italy National Cluster, Milano, Italy
e-mail: massimiliano.boggetti@clusteralisei.it

© The Author(s) 2025
T. Ferrante and M. Sacco (eds.), *Habitable Future*,
SpringerBriefs in Applied Sciences and Technology,
https://doi.org/10.1007/978-3-031-95735-2_2

2 Technological Advancements in Life Sciences

Technological advancements in the life sciences sector are both complex and captivating. They lead to early diagnoses, improved treatment options, and ultimately, a longer, healthier life for individuals. The interdisciplinary nature of this field integrates a wide array of disciplines such as chemistry, biochemistry, biotechnology, genetic engineering, electronics, robotics, physics, materials science, mathematics, computer science, artificial intelligence, and the Internet of Things—IoT.

For instance, AI-driven algorithms are increasingly utilized to analyse medical data, enhancing diagnostic accuracy and enabling timely interventions [1]. Furthermore, advancements in genetic engineering, such as CRISPR technology, have revolutionized the approach to treating genetic disorders by allowing precise modifications to DNA [2].

3 Collaborative Hubs and Open Innovation

Research driving innovation in the life sciences often flourishes within collaborative hubs or clusters. These environments foster partnerships among various stakeholders, including academic institutions, research organizations, and private companies, enabling the collective expertise necessary to develop pharmaceuticals and medical devices through a model of "open innovation".

One notable example is the ALISEI cluster in Italy, which focuses on life sciences and biotechnology. ALISEI promotes collaboration among universities, research institutions, and businesses in the region, facilitating knowledge transfer and technology commercialization. By providing a platform for networking, shared resources, and joint research initiatives, ALISEI enhances the region's capacity for innovation and accelerates the development of cutting-edge solutions in healthcare.

4 Investment in Research, Innovation, and Global Landscape

The private life sciences sector invests approximately 6% of its revenues in research and innovation, one of the highest rates among all sectors. This significant investment, together with the Public one, underscores the commitment to advancing knowledge and developing effective solutions within the field. The return on investment in life sciences research is substantial, contributing not only to improved health outcomes but also to economic growth through job creation and the development of new markets.

Investing in healthcare is particularly critical in developed countries, where innovation serves as a pathway to creating more sustainable healthcare systems. Current demographic trends are placing immense pressure on these systems. The traditional demographic pyramid, where three children supported two parents and one grandparent in the 1980s, has reversed. Today, one child often supports two parents and three grandparents [3]. This shift is creating significant challenges for healthcare systems, which must adapt to an aging population that requires more resources and support.

The health benefits of innovation are paramount; however, access to innovative medical care is often limited to affluent regions, leaving significant populations in areas such as Africa and parts of Asia underserved. This disparity in healthcare access raises ethical concerns and underscores the importance of equitable healthcare policies.

Research in life sciences is crucial for developing simple yet effective solutions that cater to the unique challenges faced by developing countries. By focusing on cost-effective innovations, researchers can create affordable medical technologies that are accessible to underserved populations, thereby improving health outcomes and fostering equity in healthcare access. Global initiatives aimed at increasing access to essential medicines and technologies are crucial in bridging this gap and ensuring that advancements benefit all populations, regardless of geographical or economic barriers [4].

5 Importance of a Human-Centric Approach

A human-centric approach is essential in medical device design, particularly for older patients managing chronic conditions. Devices must prioritize usability, accessibility, and simplicity through intuitive designs, ergonomic features, and effective user education. Integrating telemedicine and remote monitoring empowers elderly patients, enhancing adherence, independence, and health outcomes. Engaging users directly in the design process further ensures that medical technologies meet the evolving needs and capabilities of an aging population.

6 Conclusion

Despite challenges, the future of medical innovation remains promising, driven by continued investment in research, interdisciplinary collaboration, and a commitment to equitable healthcare access globally.

Innovations in life sciences are more than mere tools: they serve as extensions of human capabilities. Thoughtfully designed and integrated into patients' lives, they unlock new skills, empowering individuals to manage their health effectively. This synergy generates a positive feedback loop, where enhanced human capacity improves health outcomes, stimulating further technological innovation.

Concepts such as the "superhuman," where technology, biotechnology and chemical substances significantly augments human abilities, illustrate the transformative potential of innovation in this field.

Ultimately, research in medicine not only improves health but may also open the door to the next stage of human evolution.

References

1. Topol, E. J. (2019). *Deep medicine: How artificial intelligence can make healthcare human again*. Basic Books.
2. Doudna, J. A., & Charpentier, E. (2014). The new frontier of genome engineering with CRISPR-Cas9. *Science, 346*(6213), 1258096. https://doi.org/10.1126/science.1258096
3. OECD. (2021). *Health at a glance: OECD indicators*. https://www.oecd.org/en/publications/health-at-a-glance-2021_ae3016b9-en/full-report.html
4. World Health Organization. (2022). *Tracking universal health coverage: 2021 global monitoring report*. https://www.who.int/publications/i/item/9789240040618

Open Access This chapter is licensed under the terms of the Creative Commons Attribution 4.0 International License (http://creativecommons.org/licenses/by/4.0/), which permits use, sharing, adaptation, distribution and reproduction in any medium or format, as long as you give appropriate credit to the original author(s) and the source, provide a link to the Creative Commons license and indicate if changes were made.

The images or other third party material in this chapter are included in the chapter's Creative Commons license, unless indicated otherwise in a credit line to the material. If material is not included in the chapter's Creative Commons license and your intended use is not permitted by statutory regulation or exceeds the permitted use, you will need to obtain permission directly from the copyright holder.

Silver Economy for Innovation and Inclusivity: Sustainable Environments for Life Independency

Felice Lopane

Abstract The Silver Economy represents a crucial economic and social transformation driven by an aging population, particularly in Europe. As life expectancy increases, so do the challenges related to frailty, chronic diseases, and social isolation. This paper explores the demographic shift, its impact on healthcare systems, and the role of assistive technologies in fostering independent and healthy aging. Smart cities, digital health tools, and home automation are emerging as key solutions to address the needs of elderly citizens, improving their quality of life while ensuring economic sustainability. Italy, at the forefront of this transition, is investing in digital healthcare solutions to enhance accessibility, particularly in underserved regions. By integrating innovation, public policies, and private sector engagement, the Silver Economy can drive inclusive growth, supporting both individuals and the broader welfare system.

Keywords Silver economy · Assistive technologies · Aging population

1 Demographic Scenario and the Rise of Frailty in an Ageing Society

The European Commission calls it Silver Economy, a value chain worth 5.7 trillion euro in 2025 according to the Commission estimates, blossoming from the continuous shift of our societies to the over 50 and over 65 population [1]. A consequence to a demographic change impacting all the wealthiest and developed countries around the world and Europe itself which is ageing like never before as a result of falling birth rates and improvement in life expectancy.

In the last 10 years EU population structure has been changing with an increasing share of the population aged 65 and over. In 2024, over 65 citizens reached 21.6% of the total EU population (449.3 million people), an increase of 2.9 percentage points

---

F. Lopane (✉)
Silver Economy Network, Milan, Italy
e-mail: felice.lopane@lombardialifesciences.it

© The Author(s) 2025
T. Ferrante and M. Sacco (eds.), *Habitable Future*,
SpringerBriefs in Applied Sciences and Technology,
https://doi.org/10.1007/978-3-031-95735-2_3

(pp) than 2014. The median age of the total EU population has been increasing from 2014 to 2024, and reached 44.7 years on January 1st 2024 (2.2 years more than 2014).[1] Within the European Scenario, Italy is at the forefront with a median age increase of 4 years to the national median age, also thanks to high life expectancy rates that kept on increasing right after the Covid-19 pandemic hit.

Italy is known for its blue zones (in Sardinia Region) and high quality of life, depending on a mix of genetics, lifestyle, natural assets and culture, making it one of the most important longevity hubs in the world. In the last 20 years [2] life expectancy increased of 2.4 years (80.7 years in 2004 vs. 83.1 years in 2023), while life expectancy in good health improved of almost 3 years just in 14 years (56.4 in 2009 vs. 59.2 in 2023). However, along with an increase in the number of citizens living longer, the numbers of elderly care demand, frailty, disability and loneliness have been rising at the same time, generating a "silver bubble" needing actions and policies to be correctly managed, with different dynamics and dimensions from the North (the wealthier and industrialized areas) to the South.

The ageing process brings with it an inevitable increase in the chronic conditions incidence and prevalence across the country. In Italy chronic conditions affect 40.5% of the population, so 24 million people live at least with one chronic condition, and 12.2 million Italians are affected by 2 pathologies simultaneously. Chronic diseases incidence is linked to the ageing process and as a matter of fact 85% of over 75 population lives with a chronic disease, and 64.3% of lives with two [3]. Living with a disease typically involves higher need in terms of health care, direct and indirect assistance, usually involving formal and informal caregivers. At the same time, many conditions may bring elderly patients to higher disability levels, due to physical impairments limiting their ability to manage daily routines and maintain a good level of social interaction, which is usually associated with better levels of healthcare [4].

Throughout the years different studies [5] have strengthened evidences around this concept and we may say that social relationships have a direct and indirect influence on health outcomes and mortality, being a key for policymakers needing to find solutions to the demographic change undermining the stability of our Welfare System. Social isolation may be destructive for people at all ages and above all for those who need more social and physical activity to stay healthy and live a healthy life, postponing the rise of physical and mental conditions affecting the individual life and the system stability.

---

[1] Eurostat, data from 2014 and 2024, 2025.

2 Healthy and Secure Living: Innovative Assistive Technologies to Support the Management of the Demographic Shift

Smart Cities, homes and living, this is the concept many communities are working on when thinking about how to mitigate the potential negative effects of a demographic structure tending to population segments beyond the age of 65.

In 2025, the Smart Home worldwide market is projected to reach 174.0 billion dollars, with a projected value of 250.6 billion dollars in 2029 (9.55% CAGR between 2025 and 2029), and these technologies will reach almost every household with a constant increase in the penetration rates within the period 2025–2029 (77.6% in 2025 vs. to 92.5% in 2029).

According to a recent report, the pace of this market has been incredible over time, given the high rate of innovation launched in the market across many areas: security, medical assistance, energy, connectivity, and entertainment [6]. Smart Appliances, Control & Security and Connectivity are the devices' clusters that represent the highest share of the market, and might have a direct impact on the healthiness of elderly citizens in need for protection, connections with caregivers and healthcare providers, as well as communication devices to let them get in touch with their peers.

In Italy the market was worth 900 million euro in 2024 (+11% than 2023) with security solutions and smart household appliances leading the market (respectively 28% and 19% of the total market value), and an overall 60% penetration, meaning that 6 people over 10 already use these appliances within their houses [7], a little down the worldwide average.

Within the global market, assistive technologies with a medical or health related scope play an important role. Among those, mobility aids, hearing aids, visual aids, communication and daily living aids are the ones accounting for over 25 billion euro worldwide [8]. The challenge is the integration of these technologies in private and social housing, as well as public places and work places, given the constant increase of senior workers. Moreover, the integration of digital health tools, such as wearables, medical apps, tele-health (screening, monitoring, assistance, rehabilitation) is playing a crucial role for International healthcare Systems, willing to respond to a higher demand of healthcare and promote home-care services, preserving effectiveness, efficacy, outcomes and economic sustainability for systems facing higher incidence and prevalence of metabolic, cardiovascular, and age-related conditions across the society.

Italy itself launched a new program of investments targeting digitalization and home care through the Next Generation EU and National Recovery and Resilience Plan, funded by the European Commission. Within the program, Italy is investing almost 4.5 billion euros in homecare services and actions, targeting elderly people, and telemedicine services, providing the structures of the Italian Healthcare Systems (NHS) the structure to start a digital revolution within the healthcare pathways, closing the gap between the hospitals, healthcare facilities and the territorial network (from GPs to patients).

The road for Italy is clear, digital innovation will be an asset for the NHS and a driver for better heath and communication with patients, especially in those areas where the offer for facilities and services is lower, such as rural areas, southern regions and islands, where the impact of chronic conditions, fragility and disability is higher. The Italian elderly population appears to be ready for the change according to an Italian survey launched in 2022 on citizens 60 years. Telemedicine, is gaining increasing relevance in the healthcare landscape and its adoption is closely linked to the digital and technological skills of the people involved in it (from professionals to citizens), but 66% of the elderly population responding to the interview declared to feel digitally prepared to use digital health apps and technologies. It is worth noting that despite differences in digital literacy 58% of the population declared to be interested in using telemedicine services, meaning the timing for health and home digital transformation is right for public administrations, policymakers, solutions providers and citizens themselves [9].

Assistive technologies, from home to healthcare tools, might be a revolutionary asset to mitigate the effects of changing demographics, supporting our countries to make their population healthier, safer and more productive, as well as to reduce the impact of physical and mental health conditions, and then the costs and loss linked to the increasing elderly population. Tool integration, education, and awareness among population, professionals, providers and public administrations represent some of the challenges to be faced by regulators and policymakers willing for a change.

In line with international recommendations and actions, such as the UN Decade of Healthy Ageing 2021–2030, we must promote age-friendly environments to make sure our communities foster the abilities and the possibilities of the older population. We need to make sure the population could age in safe and friendly environments, ensuring social interactions, connections, inclusion, independency and health. The partnership between public and private organization is the key to understand the needs and the potential of elderly citizens of today and tomorrow, representing the most important consumer and population target for the next 15 years, especially for innovators and SMEs representing over 90% of the enterprise structure for Europe and Italy.

References

1. European Commission. (2018). *The silver economy—Final report*. Publications Office of the European Union. https://publications.europa.eu/resource/cellar/2dca9276-3ec5-11e8-b5fe-01aa75ed71a1.0002.01/DOC_1
2. Silver Economy Network. (2024). *Scenario longevità*. https://www.silvereconomynetwork.it/wp-content/uploads/2023/10/Scenario-Longevita-Rapporto-2024-1.pdf
3. Cittadinanzattiva. (2024). *XXII rapporto sulle politiche della cronicità—Diritti sospesi*. https://www.sanitainformazione.it/wp-content/uploads/2024/12/xxii-rapporto-sulle-politiche-della-cronicita-diritti-sospesi-2024-rapporto.pdf
4. Cohen, S. (2004). Social relationships and health. *American Psychologist, 59*(8), 676.

5. Umberson, D., & Montez, J. K. (2010). Social relationships and health: A flashpoint for health policy. *Journal of Health and Social Behavior, 51*, S54–S66. https://doi.org/10.1177/0022146510383501
6. Statista. (2025). *Smart home—Worldwide*. https://www.statista.com/outlook/cmo/smart-home/worldwide
7. Polimi School of Management. (2024). *Osservatorio internet of things—Smart home*. https://www.osservatori.net/internet-of-things/
8. Market.us. (2025). *Personal mobility devices market*. https://market.us/report/global-personal-mobility-devices-market/
9. Lopane, F., & Annino, A. (2022). *Scenari evolutivi della longevità: Il valore della silver economy in Italia*. Silver Economy Network, Osservatorio Silver Economy.

Open Access This chapter is licensed under the terms of the Creative Commons Attribution 4.0 International License (http://creativecommons.org/licenses/by/4.0/), which permits use, sharing, adaptation, distribution and reproduction in any medium or format, as long as you give appropriate credit to the original author(s) and the source, provide a link to the Creative Commons license and indicate if changes were made.

The images or other third party material in this chapter are included in the chapter's Creative Commons license, unless indicated otherwise in a credit line to the material. If material is not included in the chapter's Creative Commons license and your intended use is not permitted by statutory regulation or exceeds the permitted use, you will need to obtain permission directly from the copyright holder.

The Needs of the Over-55 Population and Design Strategies for the Development of Gerontechnologies for Active and Healthy Ageing

Mattia Pistolesi and **Francesca Tosi**

Abstract The role of new technologies in promoting longevity is a complex and relevant issue. Gerontechnology, a field of research that combines the knowledge of gerontology with technological advances, is central to supporting ageing because it aims to improve prevention and care. Scientific literature and the latest data suggest that, in Italy, the difference in access to and use of technologies between the older and young populations is quite marked. In order to ensure that these technologies are acceptable, accessible and inclusive, it is crucial to know and consider the barriers (cognitive, physical and sensory limitations) that hinder their use by the older population. This paper explores how the Human-Centred Design approach, in particular the questionnaire, focus group, design workshop, and systematic literature review, facilitated the identification of such limitations and needs of the older population, using this information to develop design strategies for innovative and accessible gerontechnologies. Considering also Goal 3 of the 2030 Agenda, the challenge for the coming years is to design inclusive and win–win technologies to promote active and healthy ageing.

Keywords Gerontechnology · Human-centred design · Design for health · Active ageing · Healthy ageing · Older adults

1 Introduction

Today, the role of new technologies, including digital ones, in favour of longevity is a sensitive issue. The discipline of Design is called upon to respond to people's real needs and desires through the creation of physical, digital, tangible, and intangible

M. Pistolesi (✉) · F. Tosi
Ergonomics and Design Lab., Department of Architecture, University of Florence, Calenzano (Florence), Italy
e-mail: mattia.pistolesi@unifi.it

artefacts and to raise the quality of products and their usability, pleasantness of use, interaction, and inclusion, proposing new behaviours and promoting people's physical and psychological well-being.

Technologies play a crucial role in supporting ageing. For example, thanks to the development of apps, it is possible to counter cognitive decline. Thanks to social networks, people can connect without time and space limits for physical or social activities. With innovative sensors and wearable devices, it is possible to monitor one's state of health and prevent possible future health problems.

In this regard, the field of gerontechnology or gerontechnology, is a relatively recent field that integrates the medical knowledge of gerontology and advances in new technologies, mainly digital technologies, to improve the levels of prevention and care of the older population [1–5]. This sector includes medical devices, wearables, AI, robots and exoskeletons, virtual words, apps, smart TVs, smartwatches, smartphones and serious games designed to address the needs and challenges associated with ageing.

To build the foundations for acceptable, accessible, and inclusive technology, also in line with Goal 3 of the UN Agenda 2030,[1] it is necessary to pay attention to limitations and barriers that limit the adoption of technologies by the older population, such as cognitive, physical, and sensory limitations, technological familiarity limitations, trust limitations, and, finally, economic limitations.

In this paper, we describe how the Human-Centred Design (HCD) approach has enabled a clear definition of these limitations and needs of older adults and how they have been used to define person-centred gerontechnology design strategies.

2 Definition of *Older Adults*

The definition of older adults has changed significantly over time due to generational changes and the lengthening of life expectancy. Conventionally, the reference age for defining an older person is 65 [6]. It is usual to divide the over-65s into two macro-categories. The first refers to the four age groups, namely the young older adults (65–74 years), the older adults (75–84 years), the very old (85–99 years) and, finally, the centenarian. The second category refers to the state of health, level of autonomy and social role. There are those belonging to the third age who have good health, disposability, and good social integration, and those belonging to the fourth age who show physical decay, reduced autonomy, and low levels of independence.

At the 63rd National Congress of the Italian Society of Gerontology and Geriatrics, it was discussed raising the age limit for defining an older person from 65 to 75, demonstrating that today, a 65 year-old has the physical and cognitive fitness of a 45 year-old 30 years ago [7].

---

[1] https://sdgs.un.org/2030agenda.

Finally, a tool has been developed at the European level to measure active and healthy ageing in a given geographical context through policy-making, considering the over-55s [8].

Such categorisations may not be globally valid but depend on the life expectancy of individual nations and the social context of reference. Segnini [9] states that many people nowadays distinguish between demographic and biological old age: many people over 65, although older adults demo-graphically, are often indistinguishable from physically active and healthy people in their 50s or 60s.

2.1 Characteristics and Critical Factors

The scientific literature [10–17] defines the older person as an individual who, due to the dynamics attributable to the ageing process, experiences a decline in sensory (vision, hearing, touch and kinaesthetic, taste and smell), cognitive (attention, memory, processing speed, language comprehension and executive functioning) and physical (movement, speed and strength) abilities, which differs from subject to subject.

The typical changes and limitations of the older person have a direct correlation with proper interaction with new technologies. For example, with the increasing diffusion of touch screens and digital interfaces and new ways of interacting with tangible and intangible devices, these may discourage or, in other cases, exclude the older person from using them because they are unsuitable for this specific population group. Ageing reduces muscle strength and dexterity of the hands, making it difficult to use touch-screen devices or type on small interfaces, such as those of smartwatches. Reduced vision (e.g. cataracts, presbyopia) and hearing make it difficult to read information on small screens or to understand spoken instructions. With age, the ability to learn new skills, concentrate and remember new information decreases. As a result, learning to use new technologies and memorising new passages can be slow and frustrating.

For these reasons, design plays a crucial role in understanding the changes that old age has on both daily routines and modes of use, on attitudes and perceptions towards physical and digital products with positive impacts on both the ease of use and usability of these, but also on the acceptability of new solutions to promote active and healthy ageing. From both a social and an innovation point of view, it is necessary to reflect on how to ensure that technology can adapt to the older population, finding solutions to meet the needs of this population group so that the usability and safety of older adults can be promoted and help promote active and healthy ageing.

3 Technological Innovation and Ageing

We live in an age where technology plays a central role in people's daily lives, especially for the elderly, facilitating communication, socialisation, information, and access to services.

The last decades have been characterised by a focus on activities promoting active and healthy ageing as a population that ages well brings personal and collective benefits. As is well known, the effects of ageing are manifold; an increasing number of inactive or non-productive older adults weigh economically on a decreasing number of young people. However, active ageing can increase the social productivity of older adults both in the labour market and in volunteering and other activities, also supporting young people, thus limiting the expenditure on social health services and the consumption of medicines, with benefits, no doubt, for all parties.

Over the years, European and national social policies have been developed to improve populations' health and wellbeing, reduce health inequalities, strengthen public health, and ensure person-centred, universal, equitable, sustainable, and high-quality health systems.

Technological innovation is a resource to support active and healthy ageing. Digital technologies, trackers, social media, innovative medical devices, sensors, smart home devices, wearable devices, robotics and serious games are changing how a person takes care of himself. These new devices allow people to monitor themselves by increasing awareness of their health status, increasing health education, increasing doctor-patient communication while physically sitting on their couch at home, suggesting when, how much and how to walk and move, and suggesting which meal is best for us. For example, wearable devices and sensors can monitor vital parameters such as heart rate and sleep quality, allowing early detection of possible health problems [18]. Cognitive training apps can help slow cognitive decline by offering a way to keep older adults mentally active and engaged while improving overall wellbeing, an example being the two European research projects Buddy4All[2] and Remember-Me.[3] Apps and social networks are crucial to combat isolation and promote mobility so that the older population can actively participate in events and meetings [19, 20].

While it is true that welfare has a price to pay, which is determined by devolving everything to 'technology' on the one hand, the positive impacts are innumerable.

Suffice it to say that due to the ageing of the population, per capita expenditure will increase in the coming years [21], mainly due to the emergence of pathologies linked to the ageing process and the appearance of chronic diseases. In order to ensure that the healthcare system does not collapse, it is necessary to act with medium- to long-term measures to address this problem of financial sustainability. Technological innovation, in all its facets, can contribute to making healthcare expenditure more sustainable through innovative and low-cost solutions for diagnosis, treatment,

---

[2] http://www.aal-europe.eu/projects/buddy4all/.

[3] http://www.aal-europe.eu/projects/remember-me/.

and monitoring, as well as through the realisation of technological tools useful for prevention and education for health and wellbeing.

These effects are already evident in the technology innovation market. According to the latest [22], the *Disabled and elderly assistive technology*[4] could increase from 27 billion in 2024 to 47 billion in 2030. This figure suggests that the perception that the over 50s are technologically challenged or disconnected is no longer accurate. However, older adults are tech-savvy and eager to learn more [23, 24] and feel actively involved in this digital transaction phase.

The challenge in the coming years is to design inclusive technologies that benefit everyone. Today's situation, however, confronts us with another issue related to the limited relationship between new technologies and the older segment of the population. The scholarly literature [25–27] suggests what obstacles fuel the difference in access to and use of new technologies between the older adults and the adult population (grey digital divide) that could exclude a significant portion of the older adults who are unfamiliar with technology or who suffer from physical, cognitive and sensory limitations and thus make it difficult to use complex tools. These include the mistrust of older adults toward the use of new technologies, the still low technological literacy, and the aseptic nature of the human–machine relationship, which are some of the main obstacles to the full realisation of the potential offered by new technologies in terms of greater social inclusion of the older adults. Another significant technological risk concerns personal and medical information security and privacy. For example, medication management apps collect sensitive data, making older adults more vulnerable to security breaches and misuse of their information by third parties. Finally, there is concern about the growing dependence on technology daily. Although assistive robots and monitoring devices may increase the self-nomics of older adults, there is a risk that they may become overly dependent on these tools, reducing human contact and the ability to cope with problems without technological assistance.

In Italy, this gap is much more pronounced than in other European countries because, in recent years, people have begun to reflect on the processes of diffusion and adoption by older adults. In fact, in the latest report of the Digital Economy and Society Index (DESI) [28], Italy ranks 18th out of the 27 EU member states. However, suppose the progress of the last five years is considered. In that case, it is advancing at a breakneck pace regarding the country's digital transition (human capital and literacy, connectivity, integration of digital technologies and digital public services), also thanks to the fulfilment of the strategic goals of the National Recovery and Resilience Plan.

---

[4] This sector refers to a wide range of devices, tools and services designed to help people with disabilities or age-related challenges in their daily activities.

4 Physical and Digital Gerontechnologies for Active and Healthy Ageing

This paper reports the preliminary results achieved under Task 1.1, Spoke 9, of the national research program "Age-It Ageing well in an aging" society funded under the National Recovery and Resilience Plan (NRRP). The specific objectives of Task 1.1 are the definition and subsequent drafting of design strategies for the development of physical and digital interfaces aimed at promoting active and healthy aging of the over-55 population through the methodological framework of the Human-Centered Design approach [29], which bases its effectiveness on the centrality of the person during the design process. The following paragraphs describe the methods used to define the structure of design strategies.

4.1 Systematic Literature Review

A systematic literature review conducted according to the Prisma framework [30] was conducted. Platforms such as Scopus, Web of Science, and Researchgate were considered in the present study. The following search string was used: *Best practice OR guidelines AND Geotechnology OR Gerontechnology AND Medical Devices OR Smart products OR Ergonomics products OR User friendly products AND active aging OR healthy aging AND Elderly OR older adult.*

The research focused on scientific articles published in journals or proceedings of conferences and books in the areas related to ageing, technological innovation, design, and ergonomics, considering the following areas of research: Science Technology, Social Science, Engineering, Computer Science, and Arts and Humanities, published in English languages only. For this research, the time limit was not considered. However, the search was not limited to scientific articles, and a search was conducted in the grey and popular literature using the exact keywords but in an unstructured manner. This allowed a more appropriate search on gerontechnologies aimed at older adults.

This review produced 25,204 contributions that included 7562 (30%) duplicates. After removing the duplicates, 17,642 contributions remained. A total of 17,537 contributions were removed from the three researchers involved in the research program during the abstract screening process. This left 105 articles (including the 14 in grey literature). We then conducted full-text screening; each contribution was assigned a value between 1 and 2, where 1 represents high appropriate and 2 low appropriateness. Therefore, 47 (33 + 14 grey literature) contributions were considered (see Fig. 1).

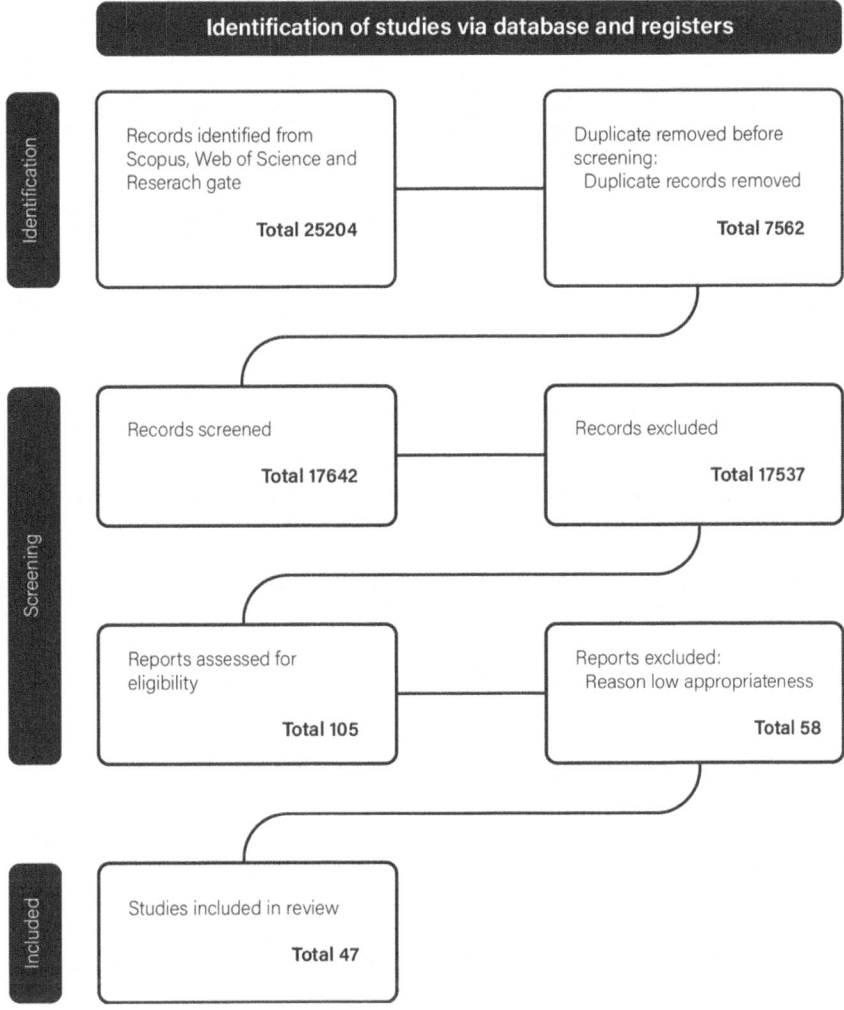

Fig. 1 PRISMA flow diagram for systematic literature review

4.2 National Questionnaire

The exploratory questionnaire aimed to understand Italian citizens' use of technologies in activities promoting and monitoring health status and well-being.

The questionnaire was developed ad hoc and consisted of dichotomous, multiple-choice, and Likert scale questions (1–5). It was administered nationwide using Facebook and LinkedIn social networks. Criteria for inclusion in the study were being

55 years of age or older, belonging to any gender, educational, and religious identity, being an Italian citizen and resident of one of the Italian regions, and, finally, possessing good cognitive skills.

The questionnaire was divided into three parts. The first part was devoted to the respondent's biographical information (marital status, gender, education level, employment, re-region and province of residence, and household composition). The second part of the State of Health about Technology consists of general questions intending to understand the health status of respondents and what activities they do to monitor and promote the state of health and well-being. The questions that make up this second part include health judgment, presence of diseases, conditions of activities of daily living that differ from ADL and IADL, and finally, mode of access to medical/health information. Finally, the last part, acceptability and experience in the use of technology, was devoted to understanding what technologies respondents use today to monitor and promote health and well-being status, what difficulties they hope for, and possible new ways of interacting with the proposed technologies. More specifically, for products such as computers, tablets, smartphones, smartwatches, mobile apps, implantable devices, smart TVs, and non-invasive medical devices, they were asked to answer the following questions: product use, years of use, problems related to the use, average daily use expressed in hours, perceived levels of experience while using with the product, sensation experienced during use, and ways of future interactions.

Due to the lack of a specific categorization defining what can be considered gerontechnology and what cannot, and paying attention to the definitions of gerontechnology that can be found in the literature, we chose to study some products that can be freely purchased on the market without restrictions and can be used even without medical training/knowledge that is used by people today for monitoring and promoting active and healthy ageing, such as personal computers, tablets, smartwatches, smartphones, smart TVs, digital assistants, AI, apps for wellness (Calm, Offtime, Forest), nutrition (Yuka, NutrInform Battery, MyFitnessPal) and sports (Nike Train-ing Club, Freeletics, Daily Yoga, etc.), wearable devices, glucometer, pulse oximeter, ECG and spirometer (Fig. 2).

The questionnaire involved 209 respondents, but only 190 met the inclusion criteria for the study.

When asked how they go about their health care needs, most (125 respondents) said they prefer to go to a general practitioner, while 57 respondents opted to seek information online. Regarding access to health services, 114 respondents prefer to go physically to a healthcare place, while 91 wish to interact directly with medical personnel. Other methods of communication, such as telephone, instant messaging, virtual meetings, and email, were less preferred, with numbers of 24, 8, and 16 respondents, respectively.

The study also analysed users' use of various technologies for health management and promoting active ageing. Among the participants, 43 use personal computers, with a majority considering it useful for monitoring health and trusting the data collected. As for smartphones, 55 respondents use them with positive results regarding usefulness, trust and understanding of data. Only 11 people use tablets, but again, the results are encouraging. 21 respondents use smartwatches, showing a

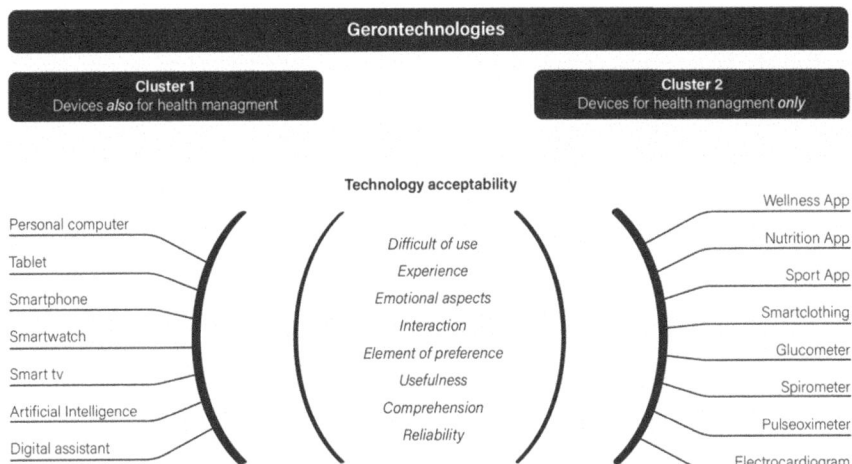

Fig. 2 Differentiation of gerontechnologies and aspects of technology acceptability investigated in the national over-55 questionnaire. On the left, cluster 1 identifies gerontechnologies also developed management and promotion of active and healthy aging, while on the right, cluster 2 identifies gerontechnologies developed only for management and promotion of active and healthy aging

good evaluation of usefulness and trust in data. Smart TV and artificial intelligence use is limited to 1 user, while only 2 respondents use digital assistants, with moderate points ratings for the factors analysed.

The study sample showed different results using technologies exclusively developed for health management. 27 people use health apps, with average positive ratings regarding usefulness, data trust, and information understanding. Only 7 respondents use food apps, and their opinion on usefulness and trust in data is low.

As for sports apps, the results are encouraging, with more than half of the users considering them valuable and trustworthy. The use of smart clothing is limited to 6 respondents, but again, the results are positive. 3 users use glucose meters, with favourable results. 27 people use the pulse oximeter, and opinions are positive regarding its usefulness and reliability. Finally, 10 users said they used the ECG, with similar results. There were no responses on the use of the spirometer. Generally, health apps and monitoring devices show good acceptance and usefulness among respondents.

4.3 Focus Group

Another important phase of the study conducted by the authors was the execution of a focus group involving 18 citizens over 55 from the Calenzano area. The data from the questionnaire made it possible to define specific research questions (What factors and aspects motivate older adults to use digital technologies?; What are the barriers?; What are the difficulties they encounter when using digital technologies?; What are the benefits that emerge during and after using digital technologies?) posed to the sample involved in this study phase. This occasion was fundamental in order to understand what motivations and factors drive the over-55s to use digital technologies, what are the barriers (cultural, economic, environmental, ethical and safety) that do not favour the use of technologies, what difficulties are experienced when using digital technologies and instead what are the benefits that emerge after the use of digital technologies.

Among the most common problems was the size of technological devices. Some are too small, and consequently, the information on the screen, the size of the buttons, and the colours do not help to facilitate the continuous use of digital objects. Other issues relate to the perceived usability of the product because they lack invitations to use it, and finally, another critical issue relates to trust in technology. Seb-well all recognise the potential of technology; many of the participants still prefer more human contact, and this result is in line with the perception stated by the sample of users who participated in the questionnaire.

4.4 Design Workshop

The last phase of the study involved organising a pro-design workshop in which 20 young designers participated. The workshop was conducted within the Design Campus, DIDA Department of Architecture.

The objective of the workshop was to interpret the data that emerged from the systemic literature review, the results of the national over-55 questionnaire, and the data that emerged from the focus group, and, finally, to define new and future scenarios of physical and digital interfaces in line with the needs of the over-55 population.

New concepts of wearable devices, smartphones, sports apps, food and health apps and new digital assistants emerged and were discussed (see Fig. 3).

5 Conclusion and Future Development

The results obtained allowed a complete definition and structuring of design strategies based on a careful analysis of physical, cognitive and sensorial needs and requirements, thanks to the methodological framework of Human-Centred Design, to make planners, researchers of various scientific disciplines, innovators, policymakers and private companies more aware of the most suitable design choices for the older population. This framework, therefore, focuses on the over 55s and the changes and limitations brought about by the ageing process and aims to suggest innovative, accessible and inclusive design directions aimed at promoting active ageing through the support and use of technological and digital systems.

Fig. 3 Some of the results emerged during the workshop with young designers

The next steps for the researchers involved in Task 1.1 are to define the guidelines and validate them with a panel of experts from specific scientific areas such as design, engineering, and psychology (see Table 1).

Table 1 Definition and structuring of project strategies necessary for the study, design and development of gerontechnologies aimed at the over-55 user group, with the objective of promoting active and healthy ageing

Factors	Sub factors	Taxonomy of design strategies
Cognitive factors	Semantic memory	3D object
		Language
		Icons
		Feedback
		Layout
	Procedural memory	Instructions
		Input
	Working memory	Text
		Icons
		Links
		Layout
		Navigation
		Interaction
	Linguistic comprehension	Language
	Attention	Layout
		Feedback
		Button
	Information processing speed	Text
		Image
		Video
		Timing
		Feedback
	Executive functioning	Feedback
		Errors
		Navigation
		Input
	Spatial cognition	3D object
		Navigation
		Feedback
		Errors
		Layout

(continued)

Table 1 (continued)

Factors	Sub factors	Taxonomy of design strategies
Functional factors	Movements	Clicking
		Tapping
		Dragging
		Zooming
		Rotating
		Scrolling
		Writing
	Motion control	Hardware
		Layout
	Movement precision	Target
		Button
		Icons
		Feedback
Sensory factors	Vision	Typeface
		Colour
		Contrast
		Text
		Layout
		UI components
		Links
	Hearing	Sound intensity
		Sound frequency
		Speech recognition
		Audio control
	Haptic	Feedback
Systemic factors	Acceptability	Emotional support
		Motivation and engagement
		Familiarity
		Digital literacy training
		Personalization and autonomy
		Stakeholder's involvement

In conclusion, the debate in scientific and public circles underlines the urgency of an in-depth reflection on ensuring that technology adapts to an ageing population. Two central questions in the relationship between technology and longevity need to be addressed: what adaptations need to be implemented in everyday technologies to ensure usability and safety for older adults and how technology can contribute to improving the quality of life of this population group.

Living longer does not necessarily mean living a healthier, more active and independent life. Active and healthy ageing is a social challenge and an opportunity all countries share. An ethical and inclusive approach is crucial to maximise the benefits of current and future technologies. This implies that technologies should be developed with respect for the dignity and rights of older adults, ensuring access to these innovations for all, regardless of their technological skills or economic conditions. Furthermore, it is essential to involve older adults and caregivers in the design process so that solutions are truly in line with their needs.

Competing Interests This study was developed within the project funded by Next Generation EU—"*Age-It*—Ageing well in an ageing society" project (PE0000015), National Recovery and Resilience Plan (NRRP)—PE8—Mission 4, C2, Intervention 1.3.

The authors have no conflicts of interest to declare that are relevant to the content of this chapter.

References

1. Bouma, H., Fozard, J. L., Bouwhuis, D. G., & Taipale, V. (2007). Gerontechnology in perspective. *Gerontechnology, 6*(4), 190–216.
2. van Bronswijk, J. E. M. H., Bouma, H., Fozard, J. L., Kearns, W. D., Davison, G. C., & Tuan, P. C. (2009). Defining gerontechnology for R&D purposes. *Gerontechnology, 8*(1), 3–10. https://doi.org/10.4017/gt.2009.08.01.002.00
3. Chen, L. K. (2020). Gerontechnology and artificial intelligence: Better care for older people. *Archives of Gerontology and Geriatrics, 91*, 104252. https://doi.org/10.1016/j.archger.2020.104252
4. Halicka, K., & Surel, D. (2021). Gerontechnology—New opportunities in the service of older adults. *Engineering Management in Production and Services, 13*(3), 114–126. https://doi.org/10.2478/emj-2021-0025
5. Huang, G., & Oteng, S. A. (2023). Gerontechnology for better elderly care and life quality: A systematic literature review. *European Journal of Ageing, 20*–27. https://doi.org/10.1007/s10433-023-00776-9
6. Istat. (2022). *Popolazione e famiglie*. Istituto Nazionale di Statistica. https://www.istat.it/storage/ASI/2022/capitoli/C03.pdf
7. Società Italiana di Gerontologia e Geriatria. (2018). *Quando si diventa anziani?*. https://www.sigg.it/wp-content/uploads/2018/12/News_Quando-si-diventa-anziani.pdf
8. United Nations Economic Commission for Europe. (2012). *Fact sheet on active ageing*. UNECE. https://unece.org/population/publications/active-ageing-index-analytical-report
9. Segnini, D. (2018). *Terza e quarta età*. https://danielesegnini.it/terza-e-quarta-eta/
10. Czaja, S. J., Boot, W. R., Charness, N., & Rogers W. A. (2019). *Designing for older adults: Principles and creative human factors approaches*. CRC Press.
11. Amarya, S., Singh, K., & Sabharwal, M. (2018). *Ageing process and physiological changes*. InTech. https://doi.org/10.5772/intechopen.76249
12. Jiang, Y., Zhang, Y., Jin, M., Gu, Z., Pei, Y., & Meng, P. (2015). Aged-related changes in body composition and association between body composition with bone mass density by body mass index in Chinese Han men over 50-year-old. *PLoS ONE, 10*(6), e0130400. https://doi.org/10.1371/journal.pone.0130400
13. Villa-Forte, A. (2025). *Effects of aging on the musculoskeletal system*. https://www.msdmanuals.com/home/bone-joint-and-muscle-disorders/biology-of-the-musculoskeletal-system/introduction-to-the-biology-of-the-musculoskeletal-system

14. Faulkner, J. A., Larkin, L. M., Claflin, D. R., & Brooks, S. V. (2007). Age-related changes in the structure and function of skeletal muscles. *Clinical and Experimental Pharmacology and Physiology, 34*(11), 1091–1096.
15. McGowen, J., Raisz, L., Noonan, A., & Elderkin, A. (2004). *Bone health and osteoporosis: A report of the surgeon general* (pp. 69–87). United States Department of Health and Human Services.
16. Vaportzis, E., Clause, M. G., & Gow, A. J. (2017). Older adults perceptions of technology and barriers to interacting with tablet computers: A focus group study. *Frontiers in Psychology, 8*, 1687. https://doi.org/10.3389/fpsyg.2017.01687
17. Yazdani-Darki, M., Rahemi, Z., Adib-Hajbaghery, M., & Izadi-Avanji, F. S. (2020). Older adults' barriers to use technology in daily life. A qualitative study. *Nursing and Midwifery Studies, 9*(4), 229–236. https://doi.org/10.4103/nms.nms_91_19
18. Tosi, F., Cavallo, F., Pistolesi, M., Fiorini, L., Rovini, E., & Becchimanzi, C. (2021). Designing smart ring for the health of the elderly: The CloudIA project. In: N. L. Black, W. P. Neumann, & I. Noy (Eds.), *Proceedings of the 21st congress of the international ergonomics association (IEA 2021). IEA 2021. Lecture notes in networks and systems* (Vol. 220). Springer. https://doi.org/10.1007/978-3-030-74605-6_48
19. Amabile, M., Hargrave, J., Clark, G., & Simunich, J. (2019). Cities alive. In *Designing for ageing communities*. Arup, London.
20. Barbabella, F., Cela, E., Socci, M., Lucantoni, D., Zannella, M., & Principi, A. (2022). Active ageing in Italy: A systematic review of national and regional policies. *International Journal of Environmental Research and Public Health, 19*(600). https://doi.org/10.3390/ijerph19010600
21. Ministero dell'Economie e delle Finanze, Dipartimento di Ragioneria generale dello stato. (2023). *Le tendenze di medio-lungo periodo del sistema pensionistico e socio-sanitario.* https://www.rgs.mef.gov.it/_Documenti/VERSIONE-I/Attivit--i/Spesa-soci/Attivita_di_previsione_RGS/2024/Ltdmlpdspess-2024.pdf
22. Research and Market. (2024). *Disabled and elderly assistive technology market by product, end user—Global forecast 2025–2030.* 360iResearch.
23. Irving, P., & Chatterjee, A. (2013). *The longevity economy: From the elderly, a new source of economic growth.* http://www.milkeninstitute.org/publications/view/687
24. Mitzner, T. L., Boron, J. B., Fausset, C. B., Adams, A. E., Charness, N., Czaja, S. J., Dijkstra, K., Fisk, A. D., Rogers, W. A., & Sharit, J. (2010). Older adults talk technology: Technology usage and attitudes. *Computers in Human Behavior, 26*(6), 1710–1721. https://doi.org/10.1016/j.chb.2010.06.020
25. Loges, W. E., & Jung, J. Y. (2001). Exploring the digital divide: Internet connectedness and age. *Communication Research, 28*(4), 536–562.
26. van Dijk, J. A. (2006). Digital divide research, achievements and shortcomings. *Poetics, 34*(4–5), 221–235.
27. van Dijk, J. A., & Hacker, K. (2003). The digital divide as a complex and dynamic phenomenon. *The Information Society, 19*(4), 315–326.
28. European Commission. (2023). *Digital economy and society index DESI 2022, Thematic chapters.* European Commission.
29. Tosi, F., Brischetto, A., & Pistolesi, M. (2020). Human-centred design—User experience: Tools and intervention methods. In *Design for ergonomics, Springer series in design and innovation* (Vol. 2, pp. 111–128). Springer. https://doi.org/10.1007/978-3-030-33562-5_6
30. Page, M. J., McKenzie, J. E., Bossuyt, P. M., Boutron, I., Hoffmann, T. C., Mulrow, C. D., Shamseer, L., Tetzlaff, J. M., Akl, E. A., Brennan, S. E., Chou, R., Glanville, J., Grimshaw, J. M., Hróbjartsson, A., Lalu, M. M., Li, T., Loder, E. W., Mayo-Wilson, E., McDonald, S., … Moher, D. (2021). The PRISMA 2020 statement: An updated guideline for reporting systematic reviews. *BMJ, 29*(372), n71. https://doi.org/10.1136/bmj.n71

Open Access This chapter is licensed under the terms of the Creative Commons Attribution 4.0 International License (http://creativecommons.org/licenses/by/4.0/), which permits use, sharing, adaptation, distribution and reproduction in any medium or format, as long as you give appropriate credit to the original author(s) and the source, provide a link to the Creative Commons license and indicate if changes were made.

The images or other third party material in this chapter are included in the chapter's Creative Commons license, unless indicated otherwise in a credit line to the material. If material is not included in the chapter's Creative Commons license and your intended use is not permitted by statutory regulation or exceeds the permitted use, you will need to obtain permission directly from the copyright holder.

Technologies for Elderly Care: Relevance and Challenges of User Engagement in the Design and Test Phases

Lorena Rossi, **Valentina Tombolesi**, **Rachele Alessandra Marziali**, **Eleonora Bonifazi**, and **Vera Stara**

Abstract The integration of digital health technologies for elderly care has shown significant potential in enhancing quality of life, reducing social isolation, and improving healthcare efficiency. However, despite technological advancements, the widespread adoption of these innovations remains limited. The paper explores the critical role of user engagement in the design and testing phases of healthcare technologies for older adults. A key barrier lies in the mismatch between technological solutions and the actual needs, preferences, and daily realities of older users. This issue is compounded by ageism, the digital divide, and stereotypical views of older adults as a homogeneous group with limited technological capabilities. Starting from the need to better understand the relation between older adults and technology considering attitude, digital literacy and technology use we analyse the possible strategies to enhance adoption rate. The often unjustified exclusion of older adults from clinical trials, particularly in the field of digital health, creates relevant risks to undermine the validity and equity of the proposed interventions. A wider inclusion in design and testing could bridge the gap between technological innovation and user needs, fostering greater acceptance and adoption of healthcare technologies that enhance autonomy and well-being among aging populations.

Keywords Older adults · Digital health · User engagement · Digital divide

1 Introduction

There is growing evidence of the value of technologies to support older adults with both physical and mental long-term conditions, and relieve carers. Their use is becoming more widespread [1, 2], and reviews and studies are emerging that reveal that these technology-enabled services can:

L. Rossi (✉) · V. Tombolesi · R. A. Marziali · E. Bonifazi · V. Stara
IRCSS INRCA, Ancona, Italy
e-mail: l.rossi@inrca.it

- Improve the quality of life of older adults and their carers, reducing social isolation, increasing perceived health status and security, and allowing carers to balance care and work [3, 4];
- Reduce hospital admissions and patients' length of stay in hospital [5, 6];
- Benefit the broader sustainability of social and health systems and the effective use of resources, through the reductions described above [7–9].

These technologies comprehend a wide variety of services that include well established health concepts like Telecare and Telemonitoring, Smart homes, Ambient Assisted Living technology and IoT solution to support autonomous living, e-health and m-health.

Significant efforts have been dedicated in recent years, for the design and development of these technologies. However, if we look to the diffusion of these technologies in real life we see that their adoption and widespread use remain significantly lower than expected.

One of the reasons for this limited diffusion can be found in the existing mismatch between the technological solutions developed and the actual needs, preferences, and daily realities of older users [10].

Despite the involvement of end users in the design and development of new technologies is widely recognized as a crucial factor in ensuring usability, accessibility, and user satisfaction, a persistent lack of inclusion of older individuals and people with disabilities in technology research and design remains.

Too often, assistive and care technologies are conceptualized and designed primarily by researchers in technological sectors without sufficient involvement of older adults in the design and validation processes.

The lack of inclusion of older users is particularly relevant in validation studies, clinical research and randomized controlled trials where too often there are inclusion criteria with unjustified upper age limit bringing to the result that more than half of the studies approved by an ethical committee with relevance to older people do not really involve the main target users [11].

According to the Technology Acceptance Model (TAM), perceived usefulness is one of the most important predictors of user acceptance of a new technology [12].

So even if the proposed solutions may be technologically advanced, if they don't respond to practical, social, and psychological needs of their intended users the new technologies face resistance in adoption, leading to limited real-world impact.

To address this gap it is necessary a shift toward more wide use of participatory design approaches that actively involve older adults in all stages of technology development—from ideation to validation. Ensuring that assistive and care technologies are designed taking into account the everyday life experiences of aging populations, researchers and industrial developers can create solutions that are not only functional but also desirable and accessible, ultimately fostering greater acceptance and adoption.

This paper analyses the key barriers to a wider inclusion of older people in the design and evaluation of healthcare technologies diffusion proposing strategies to bridge the gap between technological innovation and user needs. By understanding and addressing these challenges, we can move toward more effective, user-centred solutions that truly enhance the well-being and autonomy of the older population.

2 Older User's Engagement in Digital Health Technology Design

2.1 Challenging Ageism

The digital divide, whether for lack of technology, digital literacy, desire, or broadband connection present in older people, has been widely recognised as a significant social and cognitive gap, influencing the spread and adoption of technology globally [13].

The negative consequences of exclusion from technology and connection to the digital world were very evident during the pandemic, especially for older adults that experience wider difficulties in accessing services.

A significant limitation in research on technology for the older adults is the tendency to consider them as a homogeneous group, characterised by cognitive decline, frailty and dependency. This stereotypical approach not only ignores the great heterogeneity of aging in terms of abilities, experiences and attitudes towards technology, but also affects the design of the technological solutions themselves. The stereotyped image of the older adults as individuals unable or uninterested in technology [14] leads to various forms of social exclusion, which manifest themselves in limited access to digital services, reduced opportunities for socialisation, and low participation in community and civic life.

Building digital technology and experiences that are accessible to older adults, a population with a higher incidence of disabilities, is critical to bridge the divide.

The exclusion of older users is evident even in scientific research. In many cases in clinical trials and randomised controlled trials, to limit the number of variables to be considered, upper age limits are applied as exclusion criteria, without any scientific justification. When researchers provide a rationale for such exclusions, it often reflects negative age bias, emphasising criteria such as the need for participants to be 'fully competent', 'reliable' or 'without cognitive impairment'. This attitude raises ethical dilemmas about the right of older people to participate in research and contribute valuable information, highlighting how age stereotypes also influence the research process.

Despite these barriers, there is growing evidence on the value of the perspectives of older people, including people with dementia or other disabilities, in both research and technology design. The direct inclusion of older people in the technology development process is crucial as they bring distinct perspectives compared to other stakeholders.

The active involvement of frail older people, including conditions such as dementia, aphasia, motor dysfunctions, ataxia, hearing or vision loss in technology design not only fosters greater adoption of proposed solutions, but also enables the real goal of technology: to mitigate frailty and allows older people to cope with the challenges of everyday life with greater autonomy [15].

Integrating older adults with disabilities into the process of co-design and testing of technologies is not only a question of inclusiveness, but has a great impact also on design effectiveness. The active participation of these users makes it possible to identify real needs and barriers to accessibility and usability that might not emerge with other user groups. For example, the adoption of touch-screen interfaces or navigation systems based on complex menus may be an obstacle for those with motor or cognitive difficulties. Similarly, technologies based on voice input may be ineffective for those with speech or hearing impairments.

In addition, the possibility of participating in the design of technological tools that meet their needs not only ensures more functional solutions, but also strengthens their sense of control over their abilities and their everyday life, improving the sense of self-efficacy and autonomy.

Finally, a truly inclusive technological design must go beyond the concept of ex post 'adaptation' and adopt a universalist perspective, in which accessibility is not a secondary addition, but a founding principle of design. This approach, known as Universal Design, involves developing solutions that can be used by people with different abilities right from the initial design phase, without the need for subsequent modifications. Adopting an inclusive model not only broadens the potential user base, but also improves the user experience for everyone, regardless of age or physical and cognitive abilities.

2.2 Breaking the Vision of Older Adults as a Monolithic Group

An important step towards greater inclusion of the older adults in the design of digital health solution requires overcoming the stereotypical view of them as a uniform whole in terms of technology use [16]. As with any other characteristic, like health status or social condition, it is crucial to take into account the huge differences between individuals with respect to their attitude towards the use of technology [17].

A significant attempt to go behind the stereotype of older adults as a homogeneous group of technology 'non-users' has been done in research from Quan-Haase and colleagues [18]. Their research tries to put together different characterizations of population like device ownership, internet use, digital skills and online activities to define a set of technology use profiles, highlighting how older adults' approach to digital media is much more complex and diverse than is often believed.

This classification in 6 clusters could be useful in the process of understanding the different barriers and motivations that influence technology adoption among older people, and consequently, in developing more effective inclusion strategies (Table 1).

Table 1 Clustering of older adults with regard to attitude to technology

Cluster	Digital skills	Online activities	Estimated % of population (%)
No users	No digital skills	No online activities	10
Reluctant	Low digital skills	0–2 online activities	17
Apprehensive	Low digital skills	3+ online activities	17
Basic users	Mid-level digital skills	1–2 online activities	27
Go-getters	Mid-level digital skills	3+ online activities	22
Savvy users	High digital skills,	3+ online activities	7

The first cluster is represented by 'No users'. They report the absence of digital skills and not engagement in any kind of online activity. They fear that learning to use new technologies is too difficult for them to make the task worthwhile.

The elderly defined as 'Reluctant' represent another group distant from the digital world. In general, these individuals have a strong resistance to technology. They may have some basic digital skills and sometimes engage in some online activities but they consider the technology alien to their lifestyle, overly complex or lacking tangible added value.

'Apprehensive users', on the other hand, despite having very limited digital skills, show a greater willingness to experiment than reluctant ones. They attribute their low skill level to a lack of exposure earlier in their lives, either at work or school, and they often experience anxiety using digital media. Yet despite their fears, this group integrates some online activities, such as email, Skype, and online search, into their routines.

'Basic Users' are those who possess medium-level digital skills and use technology mainly for simple, functional tasks. They feel more comfortable than the Apprehensives, even though their use of digital media is typically more. They consider their digital skills weak compared to younger relatives, but good enough in comparison to their peers.

'Go-Getters' are distinguished from basic users by their greater involvement in digital activities. Although they do not see themselves as experts, their curiosity toward digital media brings them to be more open to new experiences. These individuals are not limited to the essential use of technology, but begin to experiment with more complex tools, such as social media, home banking or telemedicine.

'Savvy Users' report to have both a high level of digital skill and a frequent engagement in a wide range of online activities. They feel confident in their skills and do not have the anxiety most other older adults experienced. This category includes individuals who not only use technology fluently, but often act as a reference for other seniors, becoming digital ambassadors within their communities.

As with any classification, this grouping represents a simplification of a complex problem, but underlines the importance of developing technological solutions that take into account the diversity of older people as digital users, avoiding standardised approaches that risk excluding a significant part of this population.

In particular:

- For reluctant and apprehensive users, it is crucial to reduce the complexity of interaction with technology by providing simple and intuitive tools based on accessible interfaces. Moreover, training and support can be decisive in encouraging adoption.
- For basic and active users, the challenge is to increase their confidence and involvement by offering tools that progressively adapt to their needs and skills. The integration of customisation and assistance features (e.g. interactive tutorials or virtual assistants) could facilitate their use.
- For experienced users, the technology should offer opportunities for in-depth learning and active involvement, enabling them to perform advanced activities and even support other older people in adopting the technology.

3 Increasing the Presence of the Older Adults in the Evaluation of Digital Health Solutions to Assess Effectiveness and Usability

3.1 Issues in Participation of Older Adults in Clinical Studies

The participation of the older population in clinical trials has shown limited involvement, raising important questions about the equity and validity of the interventions developed [19]. This phenomenon has been present in the past for pharmacological studies and continues for studies related to the use of digital health technologies. The problem originates in a number of methodological practices and criteria adopted in the design of clinical trials, particularly in Randomised Controlled Trial (RCT) [20] which aim is to minimise the influence of confounding variables to ensure the integrity and reproducibility of the results. So RCTs are based on rigorous inclusion and exclusion criteria, whose purpose is to create homogeneous groups in order to reduce the variability of the data and, consequently, increase the statistical power of the outcomes obtained. This brings typically to the exclusion of the older adults, where multimorbidity is a prevalent condition, because comorbidities not only increase the risk of adverse events, but also complicate the interpretation of results, making it difficult to isolate the specific effect of the intervention under consideration.

A further element contributing to the marginalisation of the older adults in clinical studies is the requirement of being fully competent [21]. The presence of cognitive decline represents a problem both for the issue related to consensus and because the condition is often regarded as a confounding and risk factor in itself (for example for the risk of reducing compliance) and is therefore an exclusion criterion. This approach, although justified by the need to ensure the safety of the participants and the clarity of the data collected, inevitably leads to a selection that penalises an entire section of the population which, paradoxically, could benefit more from therapeutic innovations and targeted health services.

The exclusion of older adults from clinical trials is not without consequences [22, 23]. Firstly, it limits the generalisability of the results, since the data obtained do not reflect the complexity and variability of the real population. Secondly, this practice may contribute to inequality in access to innovative treatments and new therapies, with negative repercussions on the quality of life of the most vulnerable individuals. The lack of representation of the older adults in studies, in fact, raises the problem of 'clinical neglect', where scientific evidence is not sufficiently oriented towards the real needs of a significant part of the population.

In recent years, the importance of including a greater diversity of subjects in clinical trials has begun to be recognised, and strategies are being adopted to overcome the limitations imposed by traditional criteria. Among these, the adoption of more flexible study designs and the use of innovative approaches to manage multimorbidity and cognitive decline, such as participant stratification and subgroup analysis, represent a key step towards more inclusive and representative research.

3.2 Challenges in Validation and Evaluation of Digital Health Technologies

In addition to the general problems, the validation of the effectiveness and usability of Digital health technologies present unique and complex challenges due to intrinsic complexity of the interventions [24], such as multifactorality, need for long-term evaluations and presence of regulatory requirements.

The main challenges are summarized as follows:

Complexity: Digital health technologies often operate within a broader ecosystem of healthcare services, requiring validation that includes interactions with external systems and providers. For example, a platform that monitors vital signs and alerts for caregivers must be evaluated not only on its accuracy, but also on its integration and compatibility with hospital and caregiver workflows [25]. The effectiveness of these technologies is thus influenced by factors outside the technology itself, such as healthcare provider responsiveness, data interoperability, and alignment with medical protocols.

Long-Term Effectiveness and Sustainability: Digital health technologies for the care of older adults are often intended for continuous, long-term use. This is particularly relevant when the use is associate to prevention or management of long term conditions [26]. Assessing effectiveness over time involves evaluating how these technologies adapt to users' evolving health needs and whether they sustain user engagement and beneficial health outcomes. Conducting longitudinal studies to track effectiveness across months or years, however, is both costly and complex. Participant attrition and changing health conditions further complicate assessments, often requiring adaptive study designs that can adjust to these variations.

Customizability and Adaptability: Digital health systems are often expected to adapt to users' health needs change. Technologies that support customization, such as adjusting alert thresholds or modifying device sensitivity, can enhance usability by accommodating these changes. However, the customization and adaptation can increase complexity, reducing the comparability of the results.

Diverse User Populations: One of the most prominent usability challenges is addressing the diverse user populations that digital health technologies aim to serve. These users include older adults with varied levels of digital literacy, healthcare providers with limited time for extensive training, and caregivers who may have competing demands on their attention. Digital Health systems need to be accessible and usable for individuals with sensory impairments, cognitive decline, and physical disabilities, adding multiple layers of complexity.

Regulatory frameworks: Digital health technologies are subject to different regulatory frameworks depending on their intended use and geographic deployment. In the U.S., for instance, the FDA (Food and Drug Administration) has specific guidelines for medical devices, while in the European Union the MDR (Medical Device Regulation) entered into application on 26 May 2021 is still in the transition period.

3.3 Simplify and Improve Older Adults' Participation to the Testing of Digital Health Technologies

For the older population, already historically under-represented in clinical trials, the difficulties are accentuated when we consider trials aimed at testing the usability and effectiveness of digital health technologies. In a real-world context, digital interventions are dynamic and constantly evolving and algorithms, user interfaces and functionalities are constantly being updated based on feedback received.

To address these challenges and foster greater participation of older people in clinical trials of digital technologies, one of the possible strategies is the use of alternative evaluation methods.

The adoption of adaptive trial designs [27, 28] represents a promising strategy. Such designs allow pre-planned modifications based on the analysis of interim data, making it possible to adjust elements such as sample size or characteristics of intervention groups in response to emerging needs. This approach not only improves trial efficiency, but also allows for a more flexible inclusion of older adults by adapting criteria to account for variability related to multimorbidities and different levels of cognitive function.

Next to RCT studies, quasi-experimental methods [29, 30] offer further advantages. These designs, such as pre-post studies, interrupted time series and cohort matching, allow digital interventions to be evaluated in real-life settings, partially overcoming the limitations of randomisation. Although such methods may be subject to bias due to confounding variables, the application of robust statistical techniques can help ensure the reliability of the results, thus favouring the inclusion of older subjects that reflect the complexity of the real population.

Finally, a mixed approach [31, 32] combining quantitative and qualitative methods is crucial for a comprehensive evaluation. While quantitative metrics—such as analysis of clinical outcomes, technology use, and cost-effectiveness assessment—offer objective data, qualitative surveys—through interviews and focus groups—allow to capture the subjective experiences of the older adults, highlighting barriers to use, usability issues, and contextual factors that may influence the adoption of new digital tools.

This methodological integration ensures a holistic view of the impact of digital technologies, helping to identify the adjustments needed to make them more accessible and relevant for an older population.

3.4 Challenges with Cognitive Impairment

Older people suffering from cognitive decline represent a population that could particularly benefit from the use of digital technologies in healthcare. Such tools allow real-time monitoring of clinical status, facilitate timely and personalised interventions and potentially could improve quality of life. However, paradoxically, this category is frequently excluded from research activities [33] and clinical trials that evaluate their effectiveness, generating a significant gap in the scientific evidence devoted to them [34].

One of the main critical issues of the involvement in research of people with cognitive impairment concern the collection of informed consent [35]. In individuals with cognitive decline, the process of obtaining consent presents not only practical problems, but also ethical and legal implications. The complexity of information documents and technical language are often inadequate to ensure complete understanding, putting the person's decision-making autonomy at risk.

Operationally, the involvement of people with cognitive impairments is further complicated by difficulties in learning and using digital technologies. The autonomous adoption of cognitively demanding devices can be burdensome for those with reduced residual abilities, negatively affecting both perception and actual use of the tools. The resulting cognitive overload not only limits technology adoption, but also exposes individuals to a greater risk of errors in use, thus increasing potential safety risks [36].

To facilitate the inclusion of this category in the development and evaluation of digital health technologies, it is essential to implement a number of strategies aimed at reducing the critical issues currently highlighted. First, it is crucial to simplify and personalise informed consent [37]. The adoption of multimedia materials, such as videos and infographics, coupled with repeated explanation sessions can significantly improve the understanding of content by older people with cognitive decline.

Furthermore, the design of particularly intuitive and accessible user interfaces [38] is imperative to facilitate the integration of digital devices into the daily routine of the older adults. Such interfaces must be designed to reduce cognitive load and make interaction with the device simple and effective. At the same time, the definition of customised training programmes and continuous assistance [39] can facilitate the learning process, improving the user experience and promoting greater autonomy.

Finally, it is necessary to implement enhanced security measures that take into account the specific needs of vulnerable users [40]. Fail-safe systems and robust data protection protocols are crucial to ensure maximum security and privacy protection for older people.

Simplified interfaces, clear language, personalized training, and fail-safe systems should be in general essential guidelines for designing devices for the older adults, but, when the target population is particularly vulnerable, the critical importance of these design principles is significantly amplified, ensuring both usability and safety.

4 Conclusions

Healthcare technologies hold immense potential to improve the lives of older adults by addressing physical and mental health challenges, reducing social isolation, and supporting caregivers. However, their adoption remains significantly lower than expected due to barriers such as ageism, digital divide, and exclusionary practices in research and design. The objective of this is paper is to highlighted these challenges while emphasizing actionable strategies to overcome them.

Too often, assistive technologies are designed without sufficient involvement from older adults themselves. This results in solutions that fail to align with their practical needs or preferences. Participatory design methods offer a promising way forward by integrating older users into every stage, from ideation to validation. This not only ensures that the technology meets their actual needs but also fosters a sense of ownership and trust, which are critical for widespread acceptance. One of the relevant critical barrier that reflects a prejudice based on ageism is the stereotypical portrayal of older adults as a homogeneous group characterized by frailty or technological incompetence. Research reveals a diverse spectrum of attitudes toward technology among older individuals, ranging from non-users to savvy adopters. Recognizing this diversity is essential for creating tailored solutions that address varying levels of digital literacy and engagement.

Ageism also manifests in clinical research through unjustified upper age limits in trials, excluding those who could benefit most from innovations. This practice not only raises ethical concerns but also limits the generalizability of findings. Including older adults, even those with disabilities or cognitive impairments in research ensures that interventions are relevant to real-world populations while fostering inclusivity.

Particular attention should be used when the target is represented by older adults with cognitive decline, as in this case ethical issues are intertwined with practical difficulties arising from possible limitations related to their condition. It is crucial to adopt engagement strategies and consent-gathering procedures that respect their specific needs, while ensuring transparency and understanding. Furthermore, the design and evaluation of technologies targeting this vulnerable segment of the population should follow, even more than usually key principles like: intuitive and simplified interfaces, clear and accessible language, customised training programmes and fail-safe systems. Only through an inclusive and caring approach it would be possible to develop truly effective and safe solutions that can improve the quality of life of older people with cognitive impairments.

In conclusion, overcoming barriers to technology adoption among older adults demands a paradigm shift toward inclusive practices in both design and research.

By actively involving aging populations in co-design processes and addressing systemic biases, researchers and developers can create solutions that truly enhance autonomy, well-being, and quality of life for older individuals. Such efforts will not only improve healthcare outcomes but also contribute to broader societal goals of equity and sustainability in aging populations.

Competing Interests This study was developed within the project funded by Next Generation EU—"*Age-It*—Ageing well in an ageing society" project (PE0000015), National Recovery and Resilience Plan (NRRP)—PE8—Mission 4, C2, Intervention 1.3.

The authors have no conflicts of interest to declare that are relevant to the content of this chapter.

References

1. Carretero, S. (2015). *Technology-enabled services for older people living at home independently*. Publications Office of the European Union.
2. Billings, J., Carretero, S., Kagialaris, G., Mastroyiannakis, T., & Meriläinen-Porras, S. (2013). The role of information technology in long-term care for older people. *Long-term care in Europe: Improving policy and practice* (pp. 252–277). Palgrave Macmillan UK.
3. Carretero, S., Stewart, J., & Centeno, C. (2015). Information and communication technologies for informal carers and paid assistants: Benefits from micro-, meso-, and macro-levels. *European Journal of Ageing, 12*, 163–173.
4. Balasubramanian, G. V., Beaney, P., & Chambers, R. (2021). Digital personal assistants are smart ways for assistive technology to aid the health and wellbeing of patients and carers. *BMC Geriatrics, 21*, 1–10.
5. Rubbio, I., Bruccoleri, M., Pietrosi, A., & Ragonese, B. (2020). Digital health technology enhances resilient behaviour: Evidence from the ward. *International Journal of Operations & Production Management, 40*(1), 34–67.
6. Abernethy, A., Adams, L., Barrett, M., Bechtel, C., Brennan, P., Butte, A., Faulkner, J., Fontaine, E., Friedhoff, S., Halamka, J., & Howell, M. (2022). The promise of digital health: Then, now, and the future. *NAM Perspectives, 2022*, 10–31478.
7. Awad, A., Trenfield, S. J., Pollard, T. D., Ong, J. J., Elbadawi, M., McCoubrey, L. E., Goyanes, A., Gaisford, S., & Basit, A. W. (2021). Connected healthcare: Improving patient care using digital health technologies. *Advanced Drug Delivery Reviews, 178*, 113958.
8. Sheikh, A., Anderson, M., Albala, S., Casadei, B., Franklin, B. D., Richards, M., Taylor, D., Tibble, H., & Mossialos, E. (2021). Health information technology and digital innovation for national learning health and care systems. *The Lancet Digital Health, 3*(6), e383–e396.
9. Juba, O. O., Olumide, A. F., Idowu David, J., & Adekunle, K. (2024). The role of technology in enhancing domiciliary care: A strategy for reducing healthcare costs and improving safety for aged adults and carers. https://doi.org/10.2139/ssrn.5023483

10. Fischer, S. H., David, D., Crotty, B. H., Dierks, M., & Safran, C. (2014). Acceptance and use of health information technology by community-dwelling elders. *International Journal of Medical Informatics, 83*(9), 624–635.
11. Bayer, A., & Tadd, W. (2000). Unjustified exclusion of elderly people from studies submitted to research ethics committee for approval: Descriptive study. *BMJ Clinical Evidence, 321*, 992–993.
12. Davis, F. D. (1989). Perceived usefulness, perceived ease of use, and user acceptance of information technology. *MIS Quarterly, 13*(3), 319–340.
13. Mubarak, F., & Suomi, R. (2022). Elderly forgotten? Digital exclusion in the information age and the rising grey digital divide. *INQUIRY: The Journal of Health Care Organization, Provision, and Financing, 59*, 00469580221096272.
14. Wandke, H., Sengpiel, M., & Sönksen, M. (2012). Myths about older people's use of information and communication technology. *Gerontology, 58*(6), 564–570.
15. Mannheim, I., Schwartz, E., Xi, W., Buttigieg, S. C., McDonnell-Naughton, M., Wouters, E. J. M., & van Zaalen, Y. (2019). Inclusion of older adults in the research and design of digital technology. *International Journal of Environmental Research and Public Health, 16*(19), 3718.
16. Loos, E., Fernández-Ardèvol, M., Rosales, A., & Peine, A. (2022). Why it is easier to slay a dragon than to kill a myth about older people's smartphone use. *International conference on human-computer interaction* (pp. 212–223). Springer International Publishing.
17. Rampioni, M., Moșoi, A. A., Rossi, L., Moraru, S. A., Rosenberg, D., & Stara, V. (2021). A qualitative study toward technologies for active and healthy aging: A thematic analysis of perspectives among primary, secondary, and tertiary end users. *International Journal of Environmental Research and Public Health, 18*(14), 7489.
18. Quan-Haase, A., Williams, C., Kicevski, M., Elueze, I., & Wellman, B. (2018). Dividing the grey divide: Deconstructing myths about older adults' online activities, skills, and attitudes. *American Behavioral Scientist, 62*(9), 1207–1228.
19. Bugeja, G., Kumar, A., & Banerjee, A. K. (1997). Exclusion of elderly people from clinical research: A descriptive study of published reports. *BMJ, 315*(7115), 1059.
20. Florisson, S., Aagesen, E. K., Bertelsen, A. S., Nielsen, L. P., & Rosholm, J. U. (2021). Are older adults insufficiently included in clinical trials?—An umbrella review. *Basic & Clinical Pharmacology & Toxicology, 128*(2), 213–223.
21. Kopelman, L. (1986). Consent and randomized clinical trials: Are there moral or design problems? *The Journal of Medicine and Philosophy, 11*(4), 317–345.
22. van Marum, R. J. (2020). Underrepresentation of the elderly in clinical trials: Time for action. *British Journal of Clinical Pharmacology, 86*(10), 2014–2016.
23. Nguyen, D., Mika, G., & Ninh, A. (2022). Age-based exclusions in clinical trials: A review and new perspectives. *Contemporary Clinical Trials, 114*, 106683.
24. Ventura, F., Brovall, M., & Smith, F. (2022). Beyond effectiveness evaluation: Contributing to the discussion on complexity of digital health interventions with examples from cancer care. *Frontiers in Public Health, 10*, 883315.
25. Guo, C., Ashrafian, H., Ghafur, S., Fontana, G., Gardner, C., & Prime, M. (2020). Challenges for the evaluation of digital health solutions—A call for innovative evidence generation approaches. *NPJ Digital Medicine, 3*(1), 110.
26. Gentili, A., Failla, G., Melnyk, A., Puleo, V., Tanna, G. L. D., Ricciardi, W., & Cascini, F. (2022). The cost-effectiveness of digital health interventions: A systematic review of the literature. *Frontiers in Public Health, 10*, 787135.
27. Bhatt, D. L., & Mehta, C. (2016). Adaptive designs for clinical trials. *The New England Journal of Medicine, 375*(1), 65–74.
28. Skivington, K., Matthews, L., Simpson, S. A., Craig, P., Baird, J., Blazeby, J. M., Boyd, K. A., Craig, N., French, D. P., McIntosh, E., Petticrew, M., Rycroft-Malone, J., White, M., & Moore, L. (2021). A new framework for developing and evaluating complex interventions: Update of medical research council guidance. *BMJ, 374*.
29. Maciejewski, M. L. (2020). Quasi-experimental design. *Biostatistics & Epidemiology, 4*(1), 38–47.

30. Dong, Q., Liu, T., Liu, R., Yang, H., & Liu, C. (2023). Effectiveness of digital health literacy interventions in older adults: Single-arm meta-analysis. *Journal of Medical Internet Research, 25*, e48166.
31. Östlund, U., Kidd, L., Wengström, Y., & Rowa-Dewar, N. (2011). Combining qualitative and quantitative research within mixed method research designs: A methodological review. *International Journal of Nursing Studies, 48*(3), 369–383.
32. Taherdoost, H. (2022). What are different research approaches? Comprehensive review of qualitative, quantitative, and mixed method research, their applications, types, and limitations. *Journal of Management Science & Engineering Research, 5*(1), 53–63.
33. Taylor, J. S., DeMers, S. M., Vig, E. K., & Borson, S. (2012). The disappearing subject: Exclusion of people with cognitive impairment and dementia from geriatrics research. *Journal of the American Geriatrics Society, 60*(3), 413–419.
34. Wang, Y., Wu, Z., Duan, L., Liu, S., Chen, R., Sun, T., Wang, J., Zhou, J., Wang, H., & Huang, P. (2024). Digital exclusion and cognitive impairment in older people: Findings from five longitudinal studies. *BMC Geriatrics, 24*(1), 406.
35. Prusaczyk, B., Cherney, S. M., Carpenter, C. R., & DuBois, J. M. (2017). Informed consent to research with cognitively impaired adults: Transdisciplinary challenges and opportunities. *Clinical Gerontologist, 40*(1), 63–73.
36. Blok, M., van Ingen, E., de Boer, A. H., & Slootman, M. (2020). The use of information and communication technologies by older people with cognitive impairments: From barriers to benefits. *Computers in Human Behavior, 104*, 106173.
37. Ding, J., Keagan-Bull, R., & Tuffrey-Wijne, I. (2024). It is up to healthcare professionals to talk to us in a way that we can understand: Informed consent processes in people with an intellectual disability. *BMJ Quality & Safety, 33*(5), 277–279.
38. Castilla, D., Suso-Ribera, C., Zaragoza, I., Garcia-Palacios, A., & Botella, C. (2020). Designing ICTs for users with mild cognitive impairment: A usability study. *International Journal of Environmental Research and Public Health, 17*(14), 5153.
39. Choukou, M. A., Olatoye, F., Urbanowski, R., Caon, M., & Monnin, C. (2023). Digital health technology to support health care professionals and family caregivers caring for patients with cognitive impairment: Scoping review. *JMIR Mental Health, 10*, e40330.
40. Irazoki, E., Contreras-Somoza, L. M., Toribio-Guzmán, J. M., Jenaro-Río, C., Van der Roest, H., & Franco-Martín, M. A. (2020). Technologies for cognitive training and cognitive rehabilitation for people with mild cognitive impairment and dementia: A systematic review. *Frontiers in Psychology, 11*, 648.

Open Access This chapter is licensed under the terms of the Creative Commons Attribution 4.0 International License (http://creativecommons.org/licenses/by/4.0/), which permits use, sharing, adaptation, distribution and reproduction in any medium or format, as long as you give appropriate credit to the original author(s) and the source, provide a link to the Creative Commons license and indicate if changes were made.

The images or other third party material in this chapter are included in the chapter's Creative Commons license, unless indicated otherwise in a credit line to the material. If material is not included in the chapter's Creative Commons license and your intended use is not permitted by statutory regulation or exceeds the permitted use, you will need to obtain permission directly from the copyright holder.

New Technologies for an Aging Population: Trends and Opportunities

Matteo Romagnoli and Maria Luisa Mancusi

Abstract This chapter explores the relationship between population aging and innovation in the healthcare sector, focusing on the implications of an ageing population for the development of assistive technologies, pharmaceuticals, and medical devices. Using patent data from the United States, Europe, and Japan, we use a keyword-based approach to identify patents relevant for to the treatment of age-related diseases and analyze technological trends for these innovations, and for assistive technologies, over the last 20 years. Our findings reveal a notable increase in patent activity in these areas, both in absolute numbers and in relative terms. Key trends include the growing complexity and interdisciplinary nature of these innovations, shorter patent grant times for patents relative to age-related diseases, and a significant rise in the integration of digital technologies, particularly in data processing. Taken together, these trends indicate a growing scientific and commercial interest in addressing the healthcare needs of an aging population and highlight the role of market demand in shaping research and innovation in the healthcare sector.

Keywords Patents · Technological change · Population ageing · Age-related diseases · Assistive technologies · Market-size effect

1 Introduction

Over the past few decades, global demographic patterns have experienced significant transformation due to the accelerating aging of the population. United Nations projections indicate that the number of individuals aged 65 and older will more than double by 2050, increasing from 703 million in 2019 to over 1.5 billion [1]. While

M. Romagnoli (✉) · M. L. Mancusi
Università Cattolica del Sacro Cuore, Milan, Italy
e-mail: matteo.romagnoli1@unicatt.it

M. L. Mancusi
e-mail: marialuisa.mancusi@unicatt.it

older adults already constitute a substantial segment of the population in many high-income countries, similar trends are now emerging across developing and middle-income nations. This demographic evolution—driven by declining fertility rates and rising life expectancy—has intensified the need for tailored products capable of responding to the distinct requirements and preferences of older adults, which are often inadequately addressed by conventional products and services.

The healthcare sector is among the most directly affected by the global aging trend, facing increasing demand for advanced assistive technologies, pharmaceuticals, and medical devices. In response, significant progress has been made in the development of medical treatments targeting conditions that disproportionately impact older populations. Technological innovation in healthcare has become a critical strategy for addressing the challenges posed by demographic aging and for reshaping systems and services to better support an older population. Companies are also integrating robotics, artificial intelligence (AI) and machine learning (ML) algorithms into new assistive technologies [2]. These innovations not only improve safety but also empower older adults to engage in daily activities with greater ease and confidence.

This chapter reports an economic analysis aimed at exploring how innovation in the healthcare sector is evolving in response to shifting demographic patterns, particularly the aging of the global population. We analyze the evolution of assistive technologies and innovations in pharmaceuticals and medical devices targeting age-related conditions. We find evidence that firms are responding to market opportunities created by demographic shifts, as reflected in the growing number of technologies developed in this domain. Finally, we explore both the opportunities and the challenges surrounding age-related diseases and assistive technologies, with particular attention to the transformative role of artificial intelligence.

We show that computing and information technologies are playing a growing role in advancing next-generation assistive devices, as well as driving innovation in the pharmaceutical and medical device sectors geared toward aging populations.

The remainder of this chapter is organized as follows. Section 2 provides an overview of the relevant economic literature. Section 3 describes the data used in the analysis and Sect. 4 discusses the main trends and empirical evidence. Section 5 concludes the chapter.

2 The Economic Literature on Market Demand and Innovation

The idea that changes in market demand can stimulate innovation is well-established in the economic literature. The role of demand in shaping both the scale and direction of innovation has been extensively examined, beginning with the influential study by Griliches [3] on the adoption of hybrid seed corn in U.S. agriculture. His work demonstrated a clear link between technological change, the rate of adoption of new technologies, market size and innovation profitability. Building on this foundation, Schmookler [4] identified a strong association between sales volume and innovation, arguing that the causal pathway predominantly flows from increased demand to innovation activity. Pakes and Schankerman [5] employed a more structural approach to explore the relationship between Research and Development (R&D) intensity and industry-level factors, such as demand for inputs and output growth, further illustrating the critical connection between market dynamics and innovation efforts.

More recent empirical studies have examined this relationship specifically within the pharmaceutical sector, offering valuable insights into how demographic and market forces influence drug development. Acemoglu and Linn [6], for instance, show how potential market size affects the entry of innovative pharmaceutical products, such as non-generic drugs and new molecular entities. Using FDA data on drug approvals, they identified a significant positive effect of larger market size—driven by demographic factors—on innovation. Their findings show a decline in new drug entries in therapeutic areas primarily serving younger populations, and a corresponding increase in those targeting middle-aged consumers. Notably, while the link between market size and drug entry was robust, the correlation between market size and patenting activity appeared weaker, which the authors themselves define as a puzzling result.

Another significant contribution comes from Costinot et al. [7], who investigated the interplay between domestic demand and global pharmaceutical exports. Their analysis supports the home-market effect, originally proposed by Linder [8], suggesting that countries are more likely to export drugs that are highly demanded by their own populations. The study leverages exogenous demographic data to predict a country's disease burden and related drug demand, providing further evidence that internal market size can incentivize local innovation and enhance international competitiveness.

Dubois et al. [9] offer additional quantification of the relationship between expected market size and innovation output, estimating that roughly $2.5 billion in anticipated revenue is necessary to finance the development of a single new chemical entity. Their findings reinforce the significant elasticity of innovation with respect to market expectations. Similarly, Finkelstein [10] examined the role of public policy in shaping pharmaceutical R&D, focusing on vaccines and showing that policies designed to increase vaccination coverage directly influence the number of vaccine-related clinical trials. This highlights how demand, driven in this case by public health initiatives, can steer innovation priorities.

In the context of oncology, Budish et al. [11] point to market size and growth expectations as key determinants of firm-level R&D investment. They argue that the fixed length of patent protections encourages companies to pursue research with shorter development timelines and quicker returns, thereby favoring areas with large, immediate market potential over those requiring longer-term investment. Their findings suggest that structural factors in the innovation ecosystem—such as intellectual property regimes—interact with market dynamics in shaping the trajectory of pharmaceutical innovation.

Despite the new demand generated by population aging and the resulting innovation opportunities, important challenges persist. The high cost of developing new technologies and the limited affordability of these innovations—especially in low- and middle-income countries—continue to hinder equitable access. Moreover, integrating novel healthcare solutions into existing systems requires not only financial resources but also institutional coordination and infrastructure adaptation. Against this backdrop, it becomes essential to assess whether, and to what extent, firms are actively responding to demographic change by investing technologies specifically designed to meet the needs of an aging population.

3 Identifying Innovation Related to Population Ageing

We begin by tracing and analyzing the evolution of patented innovations related to assistive technologies and treatments for age-related diseases over the past thirty years.

Patents are a widely recognized proxy for innovation, offering detailed information on new technologies, including technical fields, citation networks, inventor, applicant and geographic scope. Patent data is readily accessible from major patent offices such as the United States Patent and Trademark Office (USPTO), the European Patent Office (EPO), and the Japan Patent Office (JPO). Furthermore, patents are systematically categorized using standardized systems like the International Patent Classification (IPC) and the Cooperative Patent Classification (CPC), enabling granular, technology-specific analysis. These attributes make patent data particularly well-suited for examining long-term trends in technological progress. In addition to being accessible and highly structured, patents represent outputs of the innovation process, making them useful for assessing technological impact. This distinguishes them from other metrics used to measure innovation, such as R&D expenditures, which reflect innovation inputs rather than outcomes. Nonetheless, patent data comes with important limitations. First, the value of patented inventions is very heterogeneous; with some patents contributing a lot to technological advancement, while other represents marginal improvements on exiting technologies. Second, not all innovations are patented, as firms may opt for alternative protection mechanisms such as trade secrets or copyright. To address the first limitation, our analysis concentrates on high-quality patents, which are more likely to reflect impactful technological advancements—a point we expand on in the following section. As for the second concern, our focus on the healthcare sector helps mitigate it. The pharmaceutical and medical devices industries exhibit a high propensity to patent, owing to the significant R&D investment required, their dependence on proprietary technologies, and the relatively low cost of imitation compared to the high cost of invention [12].

To conduct our analysis, we start from patent data filed at the European Patent Office (EPO), the Japan Patent Office (JPO), and the United States Patent and Trademark Office (USPTO), covering the period from 1990 to 2018. Given the considerable variation in patent quality and value, our focus is restricted to "triadic" patents—those filed in all three of these major offices. Triadic patents are widely recognized as high-value indicators of innovation, as they reflect an inventor's strategic decision to seek protection across the world's largest and most advanced markets. According to the Paris Convention, inventors have up to 12 months from the date of the initial filing to submit equivalent applications in other countries. When this happens, the group of related applications is referred to as a "patent family." Since this process comes with additional costs for the applicant, the breadth of geographic coverage is frequently used as a proxy for patent value, as it implies a higher expected return on the invention. Prior research has shown that patents filed internationally—particularly in multiple high-value markets—tend to be associated with higher commercial potential and strategic relevance [13]. Triadic patent families, in particular, are considered a robust indicator of valuable innovation, as securing protection in the U.S., Europe, and Japan suggests a deliberate effort to commercialize the invention globally [13]. This, in turn, implies that the underlying technology has wide applicability and significant market potential across diverse regional contexts.

Within the set of triadic patents, we then isolate those that pertain specifically to technologies targeting age-related diseases and assistive technologies. To identify the first category, we use a methodological approach that integrates International Patent Classification (IPC) codes with a systematic keyword-based search. Our process begins with the identification of innovations relevant to the treatment of age-related diseases. To this end, we draw on the work of Vos et al. [14], who use data from the Global Burden of Disease Study (2019) to identify the 25 leading causes of Disability-Adjusted Life Years (DALYs) among individuals aged 75 and older (see Table 1 for the list of these causes). DALYs serve as a comprehensive measure of disease burden, capturing both premature mortality and the years of healthy life lost to disability. As defined by the World Health Organization: "One DALY represents the loss of the equivalent of one year of full health".[1]

This unified metric allows for meaningful comparisons between diseases that primarily cause death and those that primarily cause disability.

Notably, we exclude two conditions—"Falls" and "Road Injuries"—from our analysis, as they are overly broad and not uniquely tied to the aging process. For each of the remaining conditions, we construct a tailored set of keywords, accounting for alternative spellings and commonly used synonyms. The keyword list was developed using a combination of authoritative sources, including the Institute for Health Metrics and Evaluation (IHME), which coordinates the Global Burden of Disease study; the National Library of Medicine, maintained by the National Center for Biotechnology Information (NCBI); and the ICD-10-CM coding system used in medical billing. Together, these sources ensure that our keyword selection is both comprehensive and medically accurate.

To identify relevant patents, we apply our keyword set to a full-text search of patent claims using the Orbis IP database. Since claims define the legal boundaries of an invention, they offer the most reliable basis for identifying patents explicitly

---

[1] See for instance here: Disability-adjusted life years (DALYs) available at https://www.who.int/data/gho/indicator-metadata-registry/imr-details/158.

Table 1 Leading causes of DALYs for age 75 and older (*source* [14])

	Cause		Cause
1	Ischemic heart disease	14	Colorectal cancer
2	Stroke	15	Blindness and vision loss
3	Chronic Obstructive Pulmonary Disease	16	Atrial fibrillation
4	Alzheimer's disease and other dementia	17	Stomach cancer
5	Diabetes	18	Prostate cancer
6	Lower respiratory infections	19	Cirrhosis
7	Lung cancer	20	Parkinson's disease
8	Falls	21	Osteoarthritis
9	Chronic kidney disease	22	Oral disorders
10	Age-related hearing loss	23	Tuberculosis
11	Hypertensive heart disease	24	Asthma
12	Diarrhoeal diseases	25	Road injuries
13	Low back pain		

targeting the diseases on our list. While patent descriptions may include broader technical background, related technologies, or speculative applications, these elements are not necessarily covered by the legal protection granted. In contrast, the claims delineate the precise scope of the invention, ensuring that the patents retrieved are directly concerned with the treatment methods, therapeutic compounds, or medical technologies of interest. To further refine the search, we incorporate the OECD's classification system for medical and pharmaceutical technologies, focusing on IPC codes A61B, A61C, A61D, A61F, A61G, A61H, A61J, A61L, A61M, A61N, and H05G. These classifications help narrow the dataset to the most relevant technological domains within healthcare and biomedical innovation and avoid false positives. The final dataset of addressing age-related diseases spans from 1990 to 2020.

For assistive technologies, we draw on the data and framework provided in the WIPO Technology Trends 2021—Assistive Technology report [2]. This resource offers a comprehensive dataset that allows to differentiate between two distinct categories of assistive innovations: "Conventional" technologies, which are more established and widely adopted, and "Emerging" technologies, which reflect newer developments incorporating advanced features such as AI, robotics, or smart interfaces. This distinction enables us to capture both the current landscape and forward-looking trends within the assistive technology sector. For instance, conventional assistive technologies include spectacles, medication dispensers, mobility devices such as walkers and prosthetic limbs and aids. These technologies have been widely adopted and refined over time and are characterized by their broad accessibility and consistent performance. Examples of emerging assistive technologies are artificial vision systems, intelligent virtual assistants capable of responding to voice and gesture commands, brain-computer interfaces (BCIs) that enable direct neural communication with external devices and assistive robots. In other words, these technologies reflect the cutting edge of innovation in this field. The dataset of assistive technology patents spans the years 1998–2016, providing insight into the development and diffusion of these innovations.

4 Trends in Innovation Related to Population Ageing

Figure 1 illustrates the annual evolution of triadic patent families addressing age-related diseases and assistive technologies. Both categories show a clear upward trajectory over time, reflecting growing interest and investment.

Notably, the number of triadic patents associated with age-related diseases significantly exceeds those for assistive technologies. This disparity likely reflects the pharmaceutical industry's long-standing reliance on patenting as a core mechanism for protecting innovation. Note that the increase in patents addressing age-related diseases is evident not only in absolute terms but also when examining the share of patent families dedicated to these diseases on the total number of triadic patent

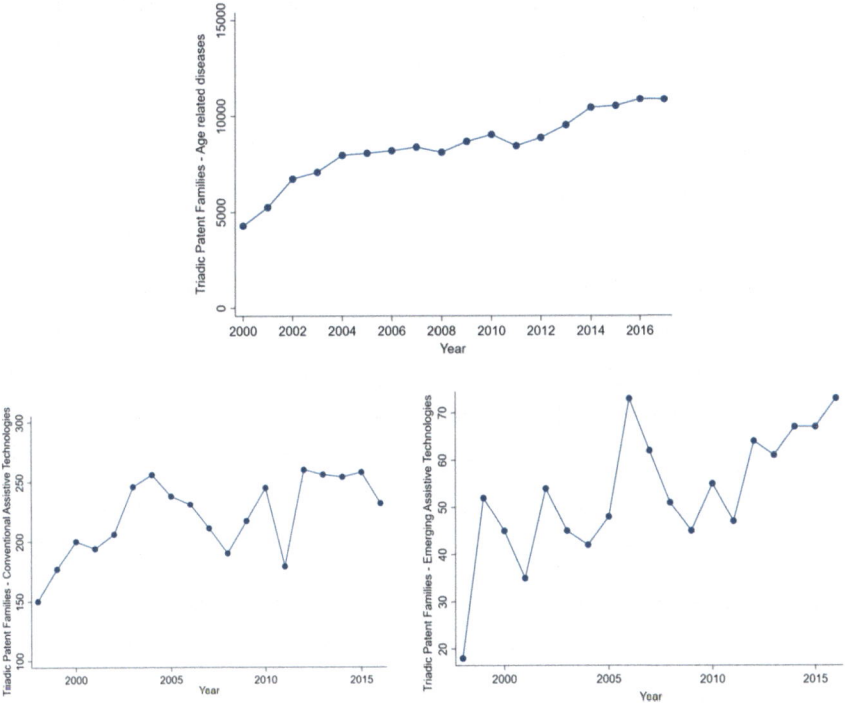

Fig. 1 Top figure: number of triadic patent families addressing age-related diseases; figures below: number of triadic patent families in conventional (left panel) and emerging (right panel) assistive technologies

families from firms in our sample (i.e. firms that patented at least one innovation related to population ageing). This share increases steadily from a little less than 2% in 2000 to roughly 4% in 2017 and highlights the growing focus on age-related health issues over time. In contrast, for assistive technologies—both conventional and emerging—there is a rise in their share up to 2010. After that point, the share declines, returning to levels roughly equivalent to those observed around the year 2000. We also look the geographical origin of innovation by examining the country of residence of the patent inventors, a common approach in the literature for linking patents to their broader development environment. The United States consistently leads in both age-related disease patents and assistive technologies. Germany ranks second in the field of age-related disease patents, followed by Japan in third place. However, the landscape shifts when examining assistive technologies as Japan overtakes Germany. Interestingly, within our triadic patent family dataset, China ranks among the top ten only in patents related to age-related diseases.

Several factors may explain this, including differing policy priorities or a focus on domestic markets rather than international patenting. Taken together, these insights offer a nuanced view of how various countries are positioning themselves to address the challenges and opportunities associated with population aging.

To better understand the characteristics of the patents in our dataset, we merge our data with the OECD Patent Quality Indicators database [13]. This resource offers a range of indicators designed to assess patent quality based on information from the patent document and the application process. We use these indicators to analyze how the nature of innovation in assistive technologies and age-related healthcare has evolved over time. Note that in this part of the analysis conventional and emerging assistive technology are analyzed together due to the small sample size.

The first indicator examined is the share of citations to scientific (non-patent) literature, which serves as a proxy for how closely an invention aligns with the scientific frontier. This indicator is calculated by dividing the number of non-patent literature citations—such as journal articles, conference proceedings, and databases—by the total number of citations in a patent [13].

In recent years, both patents addressing age-related diseases and those related to assistive technologies have shown an upward trend in their reliance on scientific literature (Fig. 2). This suggests a growing integration of cutting-edge scientific research into these innovations, and a more research-driven approach to innovation, potentially leading to more complex and breakthrough advancements. Furthermore, this result can also be interpreted as a proxy for an increase in the knowledge flow between different organizations, more specifically between firms in the private section and universities or a public research center [15]. Interestingly, this trend notably more pronounced for patents related to age-related diseases.

Overall, these patterns suggest a growing complexity of medical and healthcare innovations. As the understanding of age-related diseases evolves, the development of new treatments and technologies is increasingly driven by interdisciplinary research, integrating medical sciences with cutting-edge advancements in fields such as genetics, biology, and biotechnology.

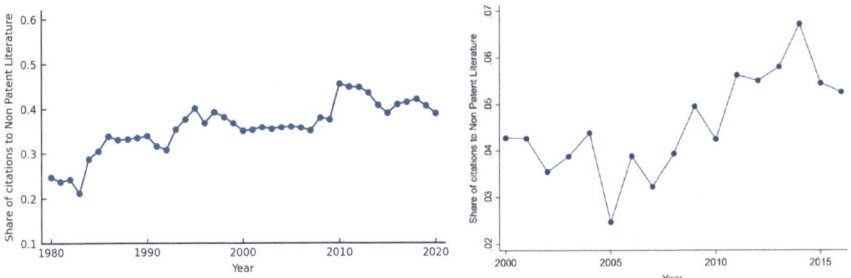

Fig. 2 Share of citations to non-patent literature for patents addressing age-related diseases (left panel) and assistive technologies (right panel)

Building on these considerations, we now examine the extent to which innovations in age-related diseases and assistive technologies incorporate knowledge from outside their core technological fields, using the Radicalness Index [13, 16].

This index captures the degree to which patents cite patents from different International Patent Classification (IPC) classes, thereby measuring the interdisciplinarity of an innovation. More precisely, the Radicalness Index is calculated by Squicciarini et al. [13] as shown in Eq. (1).

$$\text{Radicalness}_p = \sum_{j}^{n_p} CT_j / n_p; \quad IPC_{pj} \neq IPC_p \tag{1}$$

where CT_j is the number of 4-digit IPC codes of patent j (IPC_{pj}), cited by patent p, that are not assigned to the focal patent p. The denominator, n_p, is the number of total IPC classes in the backward citations of patents cited by patent p, counted at the most disaggregated level available so to normalize the index. A higher score indicates greater interdisciplinarity and, therefore, a more radical innovation.

As shown in Fig. 3, both patent categories—those related to age-related diseases sand assistive technologies—exhibit rising values of the Radicalness Index over the study period, reflecting a broader trend toward more interdisciplinary innovations. Both age-related disease technologies and assistive technologies are evolving to meet the growing challenges of complex societal needs, with both fields increasingly relying on knowledge from diverse technological and scientific areas. The cross-disciplinary nature of these technologies reflects their growing complexity and the need for integration of knowledge from multiple domains, highlighting the multifaceted nature of innovations aimed at tackling aging-related health challenges.

The final patent quality indicator we analyze is the "grant lag," which refers to the time between the filing of a patent application and its official grant date. This indicator provides insights into the perceived value of the innovation, as shorter grant lags are often associated with higher-value patents. Previous studies by Harhoff and Wagner [17] and Régibeau and Rockett [18] suggest that applicants tend to expedite

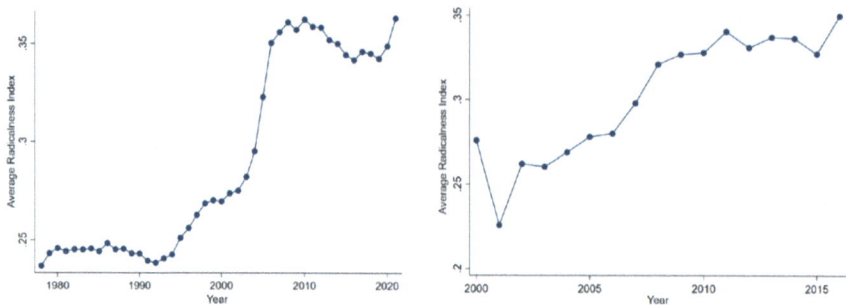

Fig. 3 Radicalness Index for triadic patent families addressing age-related diseases (left panel) and assistive technologies (right panel)

the prosecution process for patents they view as particularly valuable, potentially due to their higher commercial potential or technological significance.

Interestingly, our data reveal different trends in grant lags across the two technological domains examined—assistive technologies and age-related disease patents. While the grant lags for assistive technology patents fluctuate, but remain rather stable over the 2000–2015 period, patents addressing age-related diseases show a sharp decline in grant lags, as illustrated in Fig. 4. This difference in trends may reflect alternative dynamics within these two fields. For assistive technologies, the stable grant lags likely reflect structural factors inherent to the field—such as extended development cycles, iterative design processes, and regulatory complexities—rather than a deliberate lack of urgency on the part of applicants [2]. In contrast, the sharp decline in grant lags for age-related disease patents likely reflects the dynamic and competitive nature of this field, where rapid market entry is essential to capitalize on emerging opportunities and applicants are driven by a strong incentive to bring innovations to market swiftly.

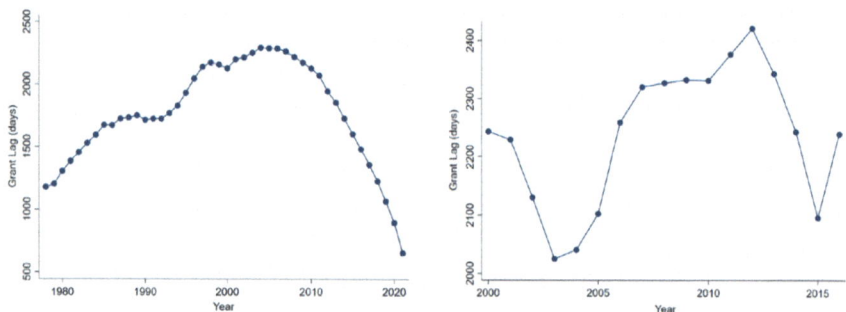

Fig. 4 Grant lag for triadic patent families addressing age-related diseases (left panel) and assistive technologies (right panel)

Finally, we examine the role of computing and artificial intelligence (AI) in the patents within our dataset by analysing the share of patents classified under technology code G06 (Computing; Calculating; Counting). This IPC code is particularly relevant for tracking technologies that related to AI, computing and data processing [19].

The results from our analysis show a marked rise in the share of patents from the G06 technology class across both assistive and age-related disease domains. This increasing integration of computing technologies highlights the expanding role of digital tools in addressing the needs of older populations and suggests that, as the aging population grows, the demand for innovative solutions to assist with health monitoring, caregiving, and disease management has spurred the adoption of computational technologies.

Figure 5 shows that, when focusing on the three sub-classes within G06 that are most relevant for AI: G06F (Electric Digital Data Processing), G06K (Data Recognition and Presentation), and G06T (Image Processing and Generation) [19] G06F stands out as the leading driver of this growth. The prominence of G06F, which deals with electric digital data processing, suggests that integration of computing in these fields is centered around improving data processing capabilities. This includes areas such as machine learning, big data analytics, and predictive modeling, which are crucial for applications in health monitoring, personalized medicine, and caregiving technologies.

This trend is consistent with the results presented before, once again suggesting that innovation targeted for the elderly is becoming more complex, research-driven, and interdisciplinary.

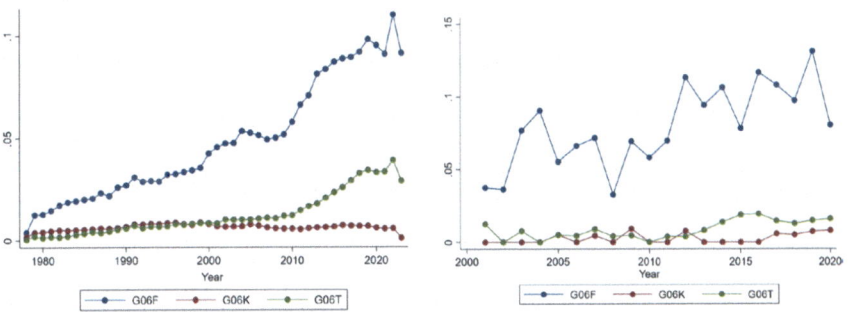

Fig. 5 Share of patents—class G06 breakdown—Patents addressing age-related diseases (left panel) and assistive technologies (right panel)

5 Conclusions

The analysis of patent data indicates a significant shift in how innovative firms respond to the challenges and opportunities of an aging population. As global demographics evolve, there is a clear trend toward leveraging technology to meet the needs of older adults. This aligns with the economic literature discussed in Sect. 2, which highlights how expanding markets attract greater investment in innovation—both in quantity and, potentially, in quality. The increasing sophistication of these innovations suggests that, beyond incentives, firms need to demonstrate a growing commitment to enhancing life quality and addressing age-related needs. Policy measures—such as funding for age-related research, regulatory support for assistive technologies, and incentives for elderly-focused innovation—may further shape the pace and direction of these developments.

Competing Interests This study was developed within the project funded by Next Generation EU—"*Age-It*—Ageing well in an ageing society" project (PE0000015), National Recovery and Resilience Plan (NRRP)—PE8—Mission 4, C2, Intervention 1.3.

The authors have no conflicts of interest to declare that are relevant to the content of this chapter.

References

1. United Nations. (2019). *World population ageing 2019*. Department of Economic and Social Affairs PD.
2. World Intellectual Property Organization. (2021). *Assistive technologies*. WIPO Technology Trends.
3. Griliches, Z. (1957). Hybrid corn: An exploration in the economics of technological change. *Econometrica, 25*, 501–522. https://doi.org/10.2307/1905380
4. Schmookler, J. (1966). *Invention and economic growth*. Harvard University Press.
5. Pakes, A., & Schankerman. M. (1984). An explanation into the determinants of research intensity. In Z. Griliches (Ed.), *R&D, patents and productivity*, University of Chicago Press.
6. Acemoglu, D., & Linn, J. (2004). Market size in innovation: Theory and evidence from the pharmaceutical industry. *The Quarterly Journal of Economics, 119*(3), 1049–1090. https://doi.org/10.1162/0033553041502144
7. Costinot, A., Donaldson, D., Kyle, M., & Williams, H. (2019). The more we die, the more we sell? A simple test of the home-market effect. *The Quarterly Journal of Economics, 134*(2), 843–894. https://doi.org/10.1093/qje/qjz003
8. Linder, S. B. (1961). *An essay on trade and transformation*. Almqvist and Wiksells.
9. Dubois, P., de Mouzon, O., Scott-Morton, F., & Seabright, P. (2015). Market size and pharmaceutical innovation. *The RAND Journal of Economics, 46*, 844–871. https://doi.org/10.1111/1756-2171.12113
10. Finkelstein, A. (2004). Static and dynamic effects of health policy: Evidence from the vaccine industry. *The Quarterly Journal of Economics, 119*(2), 527–564. https://doi.org/10.1162/0033553041382166
11. Budish, E., Roin, B. N., & Williams, H. (2015). Do firms underinvest in long-term research? Evidence from cancer clinical trials. *American Economic Review, 105*(7), 2044–2085.
12. Arundel, A., & Kabla, I. (1998). What percentage of innovations are patented? Empirical estimates for European firms. *Research policy, 27*(2), 127–141. https://doi.org/10.1016/S0048-7333(98)00033-X

13. Squicciarini, M., Dernis, H., & Criscuolo, C. (2013). Measuring patent quality: Indicators of technological and economic value. *OECD science, technology and industry working papers*, 2013/03. OECD Publishing.
14. Vos, T., Lim, S. S., Abbafati, C., Abbas, K. M., Abbasi, M., Abbasifard, M., Abbasi-Kangevari, M., Abbastabar, H., Abd-Allah, F., Abdelalim, A., Abdollahi, M., & Bhutta, Z. A. (2020). Global burden of 369 diseases and injuries in 204 countries and territories, 1990–2019: A systematic analysis for the global burden of disease study 2019. *The Lancet, 396*(10258), 1204–1222. https://doi.org/10.1016/S0140-6736(20)30925-9
15. Brusoni, S., Criscuolo, P., & Geuna, A. (2005). The knowledge bases of the world's largest pharmaceutical groups: What do patent citations to non-patent literature reveal? *Economics of Innovation and New Technology, 14*(5), 395–415.
16. Shane, S. (2001). Technological opportunities and new firm creation. *Management Science, 47*(2), 205–220. https://doi.org/10.1287/mnsc.47.2.205.9837
17. Harhoff, D., & Wagner, S. (2009). The duration of patent examination at the European Patent Office. *Management Science, 55*(12), 1969–1984. https://doi.org/10.1287/mnsc.1090.1069
18. Régibeau, P., & Rockett, K. (2010). Innovation cycles and learning at the patent office: Does the early patent get the delay? *The Journal of Industrial Economics, 58*(2), 222–246. https://doi.org/10.1111/j.1467-6451.2010.00418.x
19. Kim, J., Jun, S., Jang, D., & Park, S. (2018). Sustainable technology analysis of artificial intelligence using Bayesian and social network models. *Sustainability, 10*(1), 115. https://doi.org/10.3390/su10010115

Open Access This chapter is licensed under the terms of the Creative Commons Attribution 4.0 International License (http://creativecommons.org/licenses/by/4.0/), which permits use, sharing, adaptation, distribution and reproduction in any medium or format, as long as you give appropriate credit to the original author(s) and the source, provide a link to the Creative Commons license and indicate if changes were made.

The images or other third party material in this chapter are included in the chapter's Creative Commons license, unless indicated otherwise in a credit line to the material. If material is not included in the chapter's Creative Commons license and your intended use is not permitted by statutory regulation or exceeds the permitted use, you will need to obtain permission directly from the copyright holder.

Autonomous Devices for Elderly Monitoring and Assistance: Enhancing Quality of Life with Custom Sensors and Yottabyte-Scale Solutions

Livio D'Alvia, Christian Napoli, and Zaccaria Del Prete

Abstract Autonomous devices equipped with custom sensors represent an important advancement in elderly monitoring and assistance, offering significant potential to enhance quality of life for aging populations. In this publication, we present a comprehensive approach integrating advanced sensor technologies with yottabyte-scale data solutions, aimed at creating robust, reliable, and scalable systems for continuous monitoring and support. The proposed framework leverages state-of-the-art data processing techniques to ensure real-time responsiveness while maintaining data privacy and security. Through practical case studies and experimental setups, we demonstrate the efficacy of our system in improving mobility, health monitoring, and emergency response for elderly individuals. Our findings indicate that the integration of custom sensor networks with scalable data infrastructure can markedly reduce risks and enhance daily living conditions. This work contributes to the field of gerontechnology by providing insights into the design, implementation, and practical applications of autonomous assistance systems.

Keywords Signal processing · Remote healthcare · Personal sensor networks · Biometrics · Artificial intelligence

---

L. D'Alvia · Z. Del Prete
Department of Mechanical and Aerospace Engineering, Sapienza University of Rome, Roma, Italy
e-mail: livio.dalvia@uniroma1.it

Z. Del Prete
e-mail: zaccaria.delprete@uniroma1.it

C. Napoli (✉)
Department of Computer, Control and Management Engineering, Sapienza University of Rome, Roma, Italy
e-mail: cnapoli@diag.uniroma1.it

Institute for Systems Analysis and Computer Science, Italian National Research Council, Roma, Italy

Department of Computational Intelligence, Czestochowa University of Technology, Częstochowa, Poland

© The Author(s) 2025
T. Ferrante and M. Sacco (eds.), *Habitable Future*,
SpringerBriefs in Applied Sciences and Technology,
https://doi.org/10.1007/978-3-031-95735-2_7

1 Introduction

The rapid demographic shift towards an aging population presents significant societal and healthcare challenges. According to the World Health Organization (WHO), the proportion of people aged 60 years and older is expected to nearly double from 12 to 22% between 2015 and 2050. This demographic change increases the demand for healthcare services, long-term assistance, and monitoring to maintain an adequate quality of life for elderly individuals. Consequently, there is a growing interest in developing innovative technological solutions that can provide autonomous monitoring and assistance.

Among these innovations, autonomous devices have emerged as a viable solution to support elderly care. These devices are designed to operate independently, performing functions such as continuous health monitoring, emergency detection, and environmental control. A critical aspect of this technological advancement lies in the integration of custom sensors, which are tailored to capture specific health-related parameters, including heart rate, body temperature, movement patterns, and environmental factors like air quality and lighting conditions. These sensors are embedded in devices ranging from wearable systems to ambient intelligence setups.

The use of autonomous monitoring systems for elderly care has been extensively studied in recent years. For instance, Rashidi and Cook [1] explored the application of smart home technologies to assist elderly individuals with daily activities, emphasizing the role of ambient intelligence in maintaining safety and comfort. More recent works, such as the study by Memon et al. [2], examined the combination of wearable sensors and artificial intelligence algorithms for fall detection, highlighting the importance of data accuracy and real-time response.

Despite these advancements, challenges remain, particularly in terms of data processing, storage, and the personalization of monitoring systems.

One of the major challenges of these systems is data management. The continuous collection of data from custom sensors produces a massive volume of information, often reaching yottabyte scales when considering long-term monitoring and large populations. Efficiently processing and storing such vast amounts of data requires scalable cloud infrastructures, edge computing solutions, and robust data compression techniques. Moreover, the fusion of multimodal data from heterogeneous sensors introduces complexities in data integration and interpretation.

Furthermore, privacy and data security are fundamental issues that must be addressed to ensure the acceptance and deployment of such technologies. Autonomous devices inherently collect sensitive data, and unauthorized access or data breaches could have severe consequences. Therefore, implementing secure data transmission protocols and compliance with data protection regulations is crucial for successful adoption.

In this context, our work aims to address the gap between the technological potential of autonomous devices and the practical challenges associated with elderly care. We propose an integrated approach that combines custom sensor technologies with scalable data management solutions capable of handling yottabyte-scale data. This approach not only enhances the functionality of autonomous monitoring systems but also ensures their reliability, efficiency, and security in real-world applications.

In the following sections, we will provide a detailed analysis of the system architecture, including sensor integration, data processing techniques, and the deployment of scalable data storage solutions. Through experimental validation and real-world case studies, we will demonstrate the practical impact of our approach on improving elderly care and quality of life.

2 Related Work

The field of elderly monitoring and assistance through autonomous devices has seen significant advancements in recent years. The increasing availability of wearable technologies, smart home systems, and advanced data analytics has led to the development of integrated solutions aimed at enhancing the safety and quality of life of elderly individuals. However, despite numerous innovations, challenges related to data management, personalization, and real-time processing remain at the forefront of research.

One of the early approaches to elderly monitoring involved the use of smart home technologies. Rashidi and Cook [1] explored ambient intelligence frameworks that allow continuous monitoring of an individual's activities within a home environment. These systems often incorporate sensors embedded in the environment, such as motion detectors, pressure sensors, and smart cameras. While effective for fall detection and activity recognition, these systems face challenges related to privacy and limited data interpretation when dealing with complex health conditions.

Wearable technologies have also become an essential component of autonomous monitoring systems. Memon et al. [2] presented a framework integrating wearable sensors to detect falls and track vital signs, such as heart rate and body temperature. These systems offer portability and real-time data acquisition, making them suitable for continuous monitoring. However, their effectiveness is often limited by battery life, user compliance, and the accuracy of sensor data under varying conditions.

An innovative approach to addressing long-term psychological effects in elderly patients who have experienced COVID-19-related trauma involves the use of remote therapeutic interventions. Russo et al. [3] proposed a novel method utilizing Remote Eye Movement Desensitization and Reprocessing (EMDR) to treat traumatic disorders associated with Long-COVID and Post-COVID conditions. This approach leverages telemedicine to deliver EMDR therapy, which has shown promise in mitigating symptoms of trauma and improving psychological well-being among affected individuals. The study highlights the potential of integrating remote therapeutic practices into elderly care frameworks, especially in scenarios where in-person sessions are impractical due to health risks or mobility limitations [3]. The findings suggest that remote EMDR can be a valuable addition to the portfolio of mental health support tools in autonomous elderly monitoring systems, contributing to a more holistic care approach [3].

The development of fast and accessible eye-tracking systems has significant implications for both psychometric evaluations and human–computer interaction (HCI) applications. In [4] the authors presented an innovative neural network-based eye-tracking system designed to operate in real-time, offering a cost-effective and efficient solution for capturing eye movement data. The proposed system leverages lightweight neural network architectures to ensure rapid processing, making it suitable for applications requiring immediate feedback, such as cognitive assessments and adaptive interfaces. By reducing computational demands while maintaining high accuracy, this system addresses the limitations of traditional eye-tracking methods that often require specialized hardware or complex algorithms. The study demonstrated the potential of integrating such real-time systems into both clinical and everyday technology use, thereby enhancing accessibility and utility in various practical scenarios.

More recent studies have focused on integrating multimodal data sources to improve monitoring accuracy. For instance, Chen et al. [5] developed a system that combines environmental sensors with wearable devices to detect falls more accurately, leveraging data fusion techniques to reduce false positives. However, integrating data from heterogeneous sources poses challenges related to synchronization and consistency, especially when dealing with large datasets. The problem of data volume is further exacerbated when monitoring large populations or storing long-term data, or when using intermediary technology such as remote interfaces or virtual reality (VR). An example of the latter case has been given in [6], where the authors propose a VR-oriented study to understand how the human brain reacts to acrophobic stimuli is crucial for developing targeted interventions and therapies. In this study the authors investigate EEG patterns in young adults exposed to varying levels of acrophobia using a virtual reality (VR) environment. The research aimed to analyze the neural responses associated with different intensities of acrophobic experiences, providing insights into the cognitive and emotional processing of height-related fears. By leveraging VR to simulate real-world scenarios, the study offered a controlled and immersive setting to elicit genuine fear responses while capturing real-time brain activity. The findings indicate distinct EEG patterns corresponding to the severity of acrophobic reactions, suggesting potential biomarkers for identifying individuals at risk of severe phobic responses. This innovative approach demonstrates the value of combining VR technology with neurophysiological monitoring.

The use of scalable cloud-based architectures, as proposed by Zhang et al. [7], has enabled the processing of yottabyte-scale data. These architectures utilize edge computing to reduce latency and improve response times. However, maintaining data security and ensuring compliance with data protection regulations, such as GDPR, remains a persistent challenge. Custom sensor development has also been a critical focus, aiming to improve accuracy and efficiency. For example, Pereira et al. [8] developed custom inertial sensors for fall detection, significantly reducing power consumption compared to standard commercial devices. Customization allows for tailoring devices to specific health metrics, enhancing their relevance in elderly care. Nevertheless, integrating these custom sensors with existing platforms can be technically demanding, particularly when maintaining interoperability and data consistency.

Despite the progress made, the literature still lacks comprehensive approaches that integrate custom sensors with scalable, yottabyte-level data management systems while ensuring high data security and usability. Our work aims to address this gap by proposing a novel system architecture that combines advanced sensor technologies with robust data handling mechanisms. By leveraging real-time processing and edge computing, we aim to enhance the efficiency and reliability of elderly monitoring solutions, thereby improving their real-world applicability.

In the subsequent sections, we will describe our system architecture in detail, focusing on the integration of custom sensors, data processing pipelines, and the implementation of scalable storage solutions.

The integration of robotics and customized devices into healthcare settings, particularly during epidemic outbreaks, has proven to be an effective strategy for managing healthcare resources and supporting caregivers [9, 10].

In [11] the authors explored the use of autonomous robots as healthcare resources to alleviate the workload of medical staff, especially in critical care wards. Their study emphasizes the potential of robotic systems to perform routine monitoring, deliver medical supplies, and assist in non-invasive diagnostic procedures. By reducing direct human contact in contagious environments, these robots contribute significantly to minimizing infection risks for healthcare workers while maintaining consistent patient care. This approach underscores the importance of incorporating autonomous technologies into healthcare systems to enhance resilience during epidemics, thereby safeguarding both patients and medical personnel.

Other similar techniques are also applied on the field of personalized healthcare and rehabilitation trough the implementation of custom sensors and robotics applications [12, 13].

3 Methodology

The proposed methodology aims to design and implement autonomous devices capable of monitoring and assisting elderly individuals. The primary goal is to improve quality of life by integrating custom sensors with scalable data management solutions. The system architecture is structured to provide real-time monitoring, efficient data handling, and robust decision-making capabilities.

3.1 System Architecture

The system architecture is composed of three main components: the sensing layer, the data processing layer, and the application layer. Each layer is designed to address

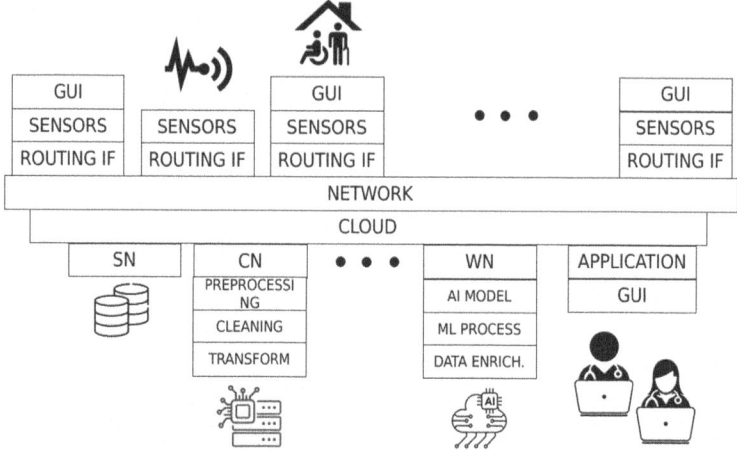

Fig. 1 A representation of the developed system to handle the asynchronous nature of the data flow, by means of time-series analysis and anomaly detection leveraging distributed computing and storage solutions to manage yottabyte-scale data

specific challenges related to data acquisition, processing, and user interaction (Fig. 1).

Sensing Layer. The sensing layer is responsible for collecting data from the elderly individuals. Custom sensors are designed to capture specific parameters relevant to health monitoring, including heart rate, body temperature, gait patterns, and environmental conditions such as room temperature and air quality. These sensors are embedded in wearable devices, stationary monitoring units, and smart home elements. The design of custom sensors prioritizes accuracy, low power consumption, and real-time data acquisition. For instance, a custom inertial sensor for fall detection should allow calibration to differentiate between sudden impacts and normal movements, reducing false positives.

Data Processing Layer. Given the large volume of data generated, efficient processing and storage mechanisms are essential. The data processing layer integrates edge computing devices to perform initial data preprocessing and filtering, minimizing latency and reducing the amount of raw data transmitted to the cloud. Real-time data aggregation techniques are employed to combine inputs from multiple sensors, allowing for the detection of critical events such as falls or irregular vital signs. Data fusion algorithms merge information from wearable and environmental sensors to increase reliability and reduce uncertainty.

For long-term data storage and large-scale analysis, the system leverages cloud-based architectures capable of handling yottabyte-scale data. A distributed database system, based on Apache Hadoop and HBase, ensures scalability and fault tolerance. Data compression algorithms, including lossless wavelet compression, are employed to minimize storage requirements without compromising data quality.

Application Layer. The application layer is the user-facing component, providing interfaces for caregivers, healthcare professionals, and elderly individuals. Mobile and web applications allow users to visualize health metrics, receive alerts, and configure monitoring settings.

Data visualization techniques are implemented to present complex data in a comprehensible format. For instance, time-series graphs display changes in vital signs, while alert dashboards summarize critical events detected by the autonomous devices. Machine learning algorithms embedded in the application layer analyze long-term trends to predict potential health deterioration, enabling proactive interventions.

3.2 Data Security and Privacy

Given the sensitive nature of health data, robust security measures are implemented. Data transmission between devices and cloud servers is secured using end-to-end encryption (AES-256), and user authentication is enforced through multi-factor methods. Additionally, data anonymization techniques are applied to protect the identity of individuals during large-scale data analysis.

3.3 System Validation and Testing

To validate the proposed methodology, a prototype system was developed and tested with a group of elderly volunteers. Performance metrics, including accuracy, latency, and power consumption, were measured under different usage scenarios. In particular, the fall detection module was tested by simulating real-life conditions, achieving an accuracy rate of 95% with a false positive rate below 2%.

Data processing efficiency was evaluated by simulating a yottabyte-scale data input using synthetic data. The system demonstrated stable performance with data retrieval times remaining under one second, even during peak loads. User feedback was collected to assess the usability of the application layer, showing a high satisfaction rate among caregivers and elderly users.

3.4 Ethical Considerations

All experimental procedures followed ethical guidelines, ensuring informed consent and data privacy. Participants were briefed on the purpose of the study, and all data was anonymized before analysis. Ethical approval was obtained from the relevant institutional review board.

4 Results and Discussion

The proposed autonomous monitoring system for elderly assistance was evaluated through a series of experimental tests conducted in both controlled environments and real-world scenarios. The evaluation aimed to assess the system's accuracy, efficiency, scalability, and user satisfaction.

4.1 System Accuracy and Reliability

The most critical aspect of the proposed system is its ability to accurately monitor health parameters and detect critical events, such as falls. The fall detection module, powered by custom inertial sensors, achieved an accuracy rate of 95% with a false positive rate of less than 2%. This was tested using both simulated falls and real-world conditions, where volunteers performed daily activities while wearing the monitoring device. Compared to conventional fall detection systems, which typically report false positive rates of around 5–10% [5], our custom sensor-based approach significantly improves reliability.

The monitoring of vital signs, including heart rate and body temperature, demonstrated high precision. Skin temperature readings exhibited a deviation of only ± 0.1 °C when compared to clinical-grade thermometers. The heart rate detection system, based on photoplethysmography (PPG), maintained an error margin of less than 2 beats per minute (BPM) when benchmarked against ECG measurements. This precision is critical for detecting early signs of health deterioration, such as fever or arrhythmias.

4.2 Data Processing and Scalability

The yottabyte-scale data management component of the system was rigorously tested to evaluate its scalability and processing speed. Using a synthetic dataset to simulate continuous monitoring from 1,000 devices over one year, the system demonstrated stable performance without significant delays. The data aggregation and fusion algorithms efficiently processed the input, maintaining response times below 500 ms even during peak data transmission periods.

Edge computing integration significantly reduced latency for real-time monitoring tasks. Compared to purely cloud-based architectures, which often exhibit latency issues due to data transmission bottlenecks (see Zhang et al. [7]), the combination of edge and cloud computing in our system reduced average response times by approximately 30%.

4.3 Data Security and Privacy

The implementation of AES-256 encryption and secure data transmission protocols ensured that no data breaches occurred during the testing phase. Data anonymization methods, particularly k-anonymity and data masking techniques, were validated through penetration testing, confirming the robustness of the system against potential cyber threats. User feedback indicated a high level of confidence in data security, which is essential for acceptance in healthcare applications.

4.4 User Experience and Satisfaction

Feedback was collected from both elderly users and caregivers during the pilot phase. Elderly users reported a positive experience with the wearable devices, citing comfort and non-intrusiveness as major advantages. The application interface was rated as intuitive, with a usability score of 8.7 out of 10 on the System Usability Scale (SUS). Caregivers appreciated the real-time alert system, particularly the visual dashboard that displayed critical events and long-term health trends.

4.5 Comparative Analysis

Compared to existing systems reviewed in the literature, our solution offers several advancements. Traditional monitoring systems often suffer from high false positive rates or limited data integration capabilities [8]. In contrast, our use of custom sensors and advanced data fusion techniques significantly reduces false alarms while providing comprehensive health monitoring. Moreover, the ability to handle yottabyte-scale data with efficient storage and processing sets our approach apart from conventional elderly monitoring frameworks.

4.6 Limitations and Future Directions

While the results are promising, some limitations were identified during testing. The accuracy of fall detection slightly decreased in highly dynamic environments where rapid movements could mimic a fall. Further calibration of the inertial sensors and the use of machine learning algorithms to better distinguish between fall-like and non-fall activities are planned.

Additionally, the energy consumption of the wearable devices, though optimized, still requires further reduction to extend battery life for continuous monitoring. Future work will focus on implementing energy-efficient algorithms and incorporating energy-harvesting techniques to enhance the system's autonomy.

4.7 Implications for Elderly Care

The proposed system represents a significant step towards improving elderly care through autonomous monitoring. By reducing false positives and enhancing real-time data processing, caregivers can make more informed decisions promptly. Furthermore, the robust data management framework supports long-term health tracking, facilitating the early detection of potential health risks. The integration of secure data handling practices ensures compliance with regulatory standards, promoting user trust and system adoption.

5 Conclusions

The aging population presents a growing challenge in modern healthcare, necessitating the development of advanced solutions to support the well-being and independence of elderly individuals. This work presented an autonomous monitoring and assistance system specifically designed for elderly care, leveraging custom sensors and scalable data management technologies to enhance quality of life.

The proposed system integrates a variety of custom sensors into wearable and ambient monitoring devices, capturing critical health metrics and environmental data in real time. By employing robust data fusion techniques and edge computing, the system effectively reduces latency while maintaining high accuracy. The integration of yottabyte-scale data management ensures that the vast amounts of information generated are processed and stored efficiently, enabling continuous monitoring and long-term health trend analysis.

Experimental results demonstrated the system's effectiveness in real-world scenarios. The fall detection module achieved an accuracy rate of 95%, significantly outperforming conventional systems that often suffer from higher false positive rates. The real-time data processing capability, facilitated by the combination of edge and cloud computing, maintained response times under 500 ms even when simulating large-scale data input. These results underscore the practical feasibility of deploying the system in everyday elderly care environments.

One of the major contributions of this work is addressing the critical challenge of handling vast data volumes generated from continuous monitoring. By integrating advanced data compression and aggregation methods, the system effectively manages yottabyte-scale data without compromising performance. Furthermore, the secure transmission protocols and data anonymization strategies implemented ensure compliance with data protection regulations, addressing a key concern in healthcare applications.

Despite these promising outcomes, several challenges remain. The system's performance in highly dynamic environments showed a slight decrease in fall detection accuracy, indicating the need for further sensor calibration and algorithm optimization. Additionally, the battery life of wearable devices still requires improvement to support long-term monitoring without frequent recharging. Future research will focus on enhancing the system's energy efficiency, exploring energy-harvesting technologies, and incorporating adaptive machine learning models to improve detection accuracy in complex scenarios.

Further validation through larger-scale pilot studies involving diverse elderly populations will be essential to generalize the findings. Collaborations with healthcare institutions and caregivers will help tailor the system to meet real-world demands, providing more personalized assistance while maintaining data integrity and privacy.

The proposed autonomous monitoring system represents a significant step forward in elderly care technology. By combining innovative sensor design, real-time processing capabilities, and robust data management, the system addresses key challenges identified in the literature. As the need for reliable and scalable elderly monitoring solutions continues to grow, the approach presented here offers a practical and forward-looking foundation for future developments in the field.

Competing Interests This study was developed within the project funded by Next Generation EU—"*Age-It*—Ageing well in an ageing society" project (PE0000015), National Recovery and Resilience Plan (NRRP)—PE8—Mission 4, C2, Intervention 1.3.

This work has been jointly developed at the *is.Lab()*—Intelligent Systems Laboratory at the Department of Computer, Control, and Management Engineering, Sapienza University of Rome, at the Thermomechanical Measurement & Biomechanics Measurement Laboratory of Department of Mechanical and Aerospace Engineering, Sapienza University of Rome, and at the Interdepartmental Laboratory of Computational Neurosciences of Sapienza University of Rome. This paper has been partially supported by the Italian Ministry of University and Research within the grant PRIN 2022 "ISIDE: Intelligent Systems for Infrastructural Diagnosis in smart-concretE", n. 2022S88WAY—CUP B53D2301318.

The authors have no conflicts of interest to declare that are relevant to the content of this chapter.

References

1. Rashidi, P., & Cook, D. J. (2009). Keeping the elderly safe: A survey on ambient intelligence and smart environments. *Journal of Ambient Intelligence and Smart Environments, 1*(2), 169–181. https://doi.org/10.3233/AIS-2009-0011
2. Memon, M., Wagner, S. R., Pedersen, C. F., Beevi, F. H. A., & Hansen, F. O. (2014). Ambient assisted living healthcare frameworks, platforms, standards, and quality attributes. *Sensors, 14*(3), 4312–4341. https://doi.org/10.3390/s140304312
3. Russo, S., Fiani, F., & Napoli, C. (2024). Remote eye movement desensitization and reprocessing treatment of long-COVID- and post-COVID-related traumatic disorders: An innovative approach. *Brain Sciences, 14*(12), 1212. https://doi.org/10.3390/brainsci14121212
4. Iacobelli E., Pelella D., Ponzi V., Russo S., & Napoli, C. (2024). A fast and accessible neural network based eye-tracking system for real-time psychometric and HCI applications. In *CEUR Workshop Proceedings* (Vol. 3870, pp. 32–41).

5. Chen, C., Jiang, X., Zhang, Z., & Sun, Y. (2021). Multimodal fall detection using data fusion of wearable and environmental sensors. *IEEE Transactions on Biomedical Engineering, 68*(2), 537–547. https://doi.org/10.1109/TBME.2020.3027893
6. Russo, S., Tibermacine, I. E., Tibermacine, A., Chebana, D., Nahili, A., Starczewscki, J., & Napoli, C. (2024). Analyzing EEG patterns in young adults exposed to different acrophobia levels: A VR study. *Frontiers in Human Neuroscience, 18*, 1348154. https://doi.org/10.3389/fnhum.2024.1348154
7. Zhang, L., Wang, J., & Li, H. (2020). Cloud-based data processing architecture for yottabyte-scale elderly monitoring. *Journal of Healthcare Engineering, 2020*, Article ID 8691342. https://doi.org/10.1155/2020/8691342
8. Zimatore, G., Cavagnaro, M., Skarzynski, P. H., Fetoni, A. R., & Hatzopoulos, S. (2020). Detection of age-related hearing losses (Arhl) via transient-evoked otoacoustic emissions. *Clinical Interventions in Aging, 15*, 927–935. https://doi.org/10.2147/CIA.S252837
9. Napoli, C., Napoli, C., Ponzi, V., Puglisi, A., Russo, S., & Tibermacine, I. E. (2024). Exploiting robots as healthcare resources for epidemics management and support caregivers. In *CEUR Workshop Proceedings* (Vol. 3686, pp. 1–10).
10. Pereira, A., Costa, M., & Silva, P. (2019). Custom inertial sensors for fall detection: Design and validation. *Journal of Sensor Technology, 9*(4), 215–223. https://doi.org/10.4236/jst.2019.94016
11. Zimatore, G., Serantoni, C., Gallotta, M. C., Guidetti, L., Maulucci, G., & De Spirito, M. (2023). Automatic detection of aerobic threshold through recurrence quantification analysis of heart rate time series. *International Journal of Environmental Research and Public Health, 20*(3), 1998. https://doi.org/10.3390/ijerph20031998
12. Avanzato, R., Mandelli, L., & Randieri, C. (2023). A NLP and YOLOv8-integrated approach for enabling visually impaired individuals to interpret their environment. In *CEUR Workshop Proceedings* (Vol. 3695, pp. 25–33).
13. Dat, N. N., Ponzi, V., Russo, S., & Vincelli, F. (2021). Supporting impaired people with a following robotic assistant by means of end-to-end visual target navigation and reinforcement learning approaches. In *CEUR Workshop Proceedings* (Vol. 3118, pp. 51–63). CEUR-WS.

Open Access This chapter is licensed under the terms of the Creative Commons Attribution 4.0 International License (http://creativecommons.org/licenses/by/4.0/), which permits use, sharing, adaptation, distribution and reproduction in any medium or format, as long as you give appropriate credit to the original author(s) and the source, provide a link to the Creative Commons license and indicate if changes were made.

The images or other third party material in this chapter are included in the chapter's Creative Commons license, unless indicated otherwise in a credit line to the material. If material is not included in the chapter's Creative Commons license and your intended use is not permitted by statutory regulation or exceeds the permitted use, you will need to obtain permission directly from the copyright holder.

Reconfiguring and Customizing Living Environments with Knowledge: The *Age-It* Decision Support System

Daniele Spoladore, Atieh Mahroo, and Marco Sacco

Abstract The aging population presents growing challenges in ensuring independent and autonomous living for older adults, particularly those with chronic conditions and disabilities. Smart environments and Ambient Assisted Living (AAL) technologies offer promising solutions by integrating intelligent systems that adapt to users' needs. However, personalizing or reconfiguring living spaces to accommodate AAL technologies requires a multidisciplinary approach that considers health conditions, assistive devices, and spatial constraints. This study presents the *Age-It* Decision Support System (DSS), a tool designed to assist architects and designers in selecting and integrating AAL solutions into domestic environments. The DSS leverages an ontological framework to model individual health conditions, environmental constraints, and assistive technologies, enabling a structured decision-making process. By incorporating a knowledge-based reasoning system, the DSS suggests user-centered and suitable modifications and device placements while ensuring compatibility with the physical space. The system communicates with Autodesk Revit, allowing seamless data exchange through XML-based file processing. This enables early-stage simulations of AAL device integration, facilitating collaborative decision-making among designers, healthcare professionals, and end users. A case study demonstrates the DSS's effectiveness in recommending assistive solutions for a user with reduced mobility, considering spatial constraints and usability factors. Future developments aim to enhance the ontology with a broader range of health conditions and assistive devices, improving its robustness and usability in real-world applications. The proposed DSS represents a significant step toward personalized and adaptive home environments, fostering independent living and improving the quality of life for older adults.

Keywords Ambient assisted living · Decision support system · Ontology · Smart environments · Building information modeling

---

D. Spoladore (✉) · A. Mahroo · M. Sacco
Institute of Intelligent Industrial Technologies and Systems for Advanced Manufacturing (STIIMA), National Research Council of Italy (CNR), Lecco, Italy
e-mail: daniele.spoladore@cnr.it

1 Introduction: Aging Population, Smart Environments, and Service Personalization

In the context of an aging population characterized by older adults, smart environments can support occupants in living independently. This section explores the main features of aging and recent technologies' role in creating smart environments. Finally, the issue of reconfiguring living environments for older adult's specific needs is addressed within the framework of the *Age-It* research project.

1.1 Aging Population

Western countries are experiencing an increase in the elderly population—i.e., citizens aged 65 and over [1]. This increase is mainly caused by the improvements in general living conditions, resulting in a higher average life expectancy. This demographic trend began emerging in the 1990s, and in some countries, including Italy, it has been accompanied by a considerable decline in birth rates. Many European countries are seeing an increase in the percentage of elderly people while the younger age groups are shrinking (the European average for older adults is around 21.3%, while in Italy, it reaches 24%) [2]. The aging phenomenon in Europe also significantly impacts healthcare services, which face increasing demands from patients with chronic illnesses [3, 4]. Moreover, it raises challenges related to managing elderly individuals (especially those with chronic and disabling conditions resulting in a physical or cognitive limitation, permanently affecting their independence): ensuring autonomous and independent living for the elderly has become increasingly urgent.

1.2 Smart Environments and Ambient Assisted Living: A Multidisciplinary Approach for Aging

The recent integration of Artificial Intelligence (AI), the Internet of Things (IoT), and distributed data management systems has driven the creation of increasingly intelligent living environments capable of offering "smart" services to their occupants. The term "smart environment" stands for a living space (equipped with devices) capable of anticipating and satisfying occupants' needs. Designing smart living environments capable of customizing their services requires a significant multidisciplinary effort [5] involving the medical sector, architects, designers, and residents: it is a process requiring understanding individual needs, identifying devices that support these needs, and designing spaces that accommodate the specific characteristics of the individual and the devices they may require.

The set of technologies (both software and hardware) operating in a living environment to monitor, support, and improve residents' quality of life was defined as

Ambient Assisted Living (AAL) [6] in the early 2000s. It quickly became apparent that AAL offered a promising solution for supporting elderly residents [7]. AAL shares four fundamental challenges with redesigning spaces for specific users; the most relevant in *Age-It* are: *a.* Identifying reliable information about the user's physical and cognitive characteristics for understanding which services to personalize; *b.* Deploying smart devices in a home, which implies context-dependent technical challenges; *c.* Designing spaces within spatial constrains (of rooms and devices)–knowledge often outside the expertise of designers and architects; *d.* Selecting and "prescribing" devices and aids must account for healthcare professionals' input.

Without a multidisciplinary approach to address these issues, designing (or reconfiguring) spaces for vulnerable residents would be challenging for architects. Therefore, within the *Age-It* project, the effort focused on designing and developing a Decision Support System (DSS) to assist architects and designers in making occupant- and AAL-oriented reconfiguration decisions. These decisions include selecting devices and aids for specific occupants and ensuring compatibility with physical environments.

The core element of the proposed DSS approach is knowledge—formalized relying on domain ontologies [8]—shared computable conceptualizations of essential information (or "facts") relevant to the domain(s) of interest. The remainder of this Chapter delves into the proposed DSS architecture (Sect. 2); Sect. 3 details the ontological framework and its engineering choices. Section 4 presents a use case to illustrate the DSS functioning. Finally, the Conclusions wrap up the main outcomes of this chapter.

2 The *Age-It* DSS for Reconfiguring Living Environments

This section outlines the architecture for the proposed DSS, focusing on its ontological framework and engineering choices.

2.1 Age-It DSS Requirements and General Architecture

Leveraging similar architectures [9, 10] and taking into account the specific requirements of architects and designers when they are called to (re)design a living environment for an elderly user, a DSS designed specifically for these specialists should take into account:

1. The tools and technologies that architects and designers use in their work practice
2. Their actual knowledge related to the domains of interest (e.g., health and disabilities of users, devices, and aids for assistance, etc.)
3. The efficiency of the timing of the support provided by the DSS

Regarding the first point, the main tool for the digital design of environments is Computer-Assisted Design (CAD), a family of software technologies for creating environment or product designs, often resulting in a three-dimensional representation (3D model) of the object. Among these, Autodesk Revit is characterized by the ability to import and export different types of files, and, for these reasons, it became one of the main players in the CAD market. Considering the diffusion and specific characteristics of Autodesk Revit, the DSS discussed below is mainly designed for this software—though, its functionalities can easily be extended to other similar design software.

Regarding the second group of requirements, architects and designers are not required to know the health condition of the user for whom they are designing a living environment: it is, therefore, necessary that the DSS takes into account that its users may be utterly agnostic regarding a good portion of the knowledge modeled in it.

Finally, the DSS results must be quickly accessible so that—upon user request—the information produced by the DSS is available. It implies the development of a software component that can receive and exchange data from Revit from the ontological layer without affecting or slowing down the users' work.

The DSS, therefore, consists of two parts (Fig. 1): 1. The ontological layer, which contains the modeled knowledge (described in the following subsection); 2. The software that allows the reception and sending of data to/from a file operable with Revit.

The DSS architecture does not act directly on the Revit software: it acquires and modifies its data. In fact, among the various options available, Revit allows export in XML, a markup language used precisely for data interchange. In this way, the DSS results can be calculated from the data entered in the XML file and reinserted into the file itself to be used in the development phase of the three-dimensional model. This "lightweight" DSS architecture has the dual advantage of allowing more agile prototypal development based on widely used data formats in practice and not directly interfering with the work of architects and designers. The inferences generated by the ontological layer are exported (in XML) by the software and added to the XML file produced by Revit, then made available to Revit file users.

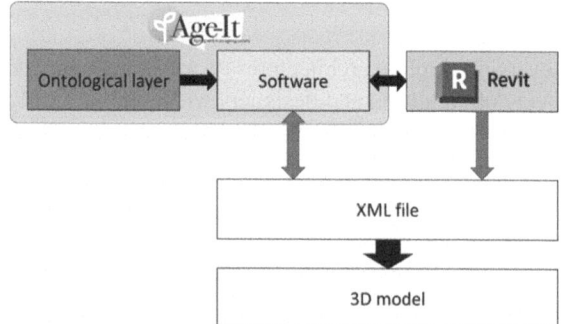

Fig. 1 A schematic representation of the *Age-It* DSS architecture. The system, composed of the ontological layer and software, acts directly on the Revit output, i.e., the XML file used to create the three-dimensional model

2.2 Ontology Engineering: Available Ontologies for Age-It

This subsection provides an overview of the main ontological models available for representing knowledge in the three areas relevant to *Age-It*—the human occupants and their health conditions, the built environment requiring reconfiguration, the modifications to be implemented in the environment.

People and health. The development of ontologies aimed at capturing relevant knowledge about individuals was among the first areas investigated in the Semantic Web. The controlled vocabulary Friend of A Friend (FOAF) [11], and Semantically-Interlinked Online Communities (SIOC) [12] are two examples of models for describing people. More recently, the Person Ontology [13] represents an ongoing effort to develop a foundational ontology about people.

Regarding the ability to model health conditions—including temporary or permanent disabilities and chronic conditions—between the early 2000s and today, there has been a shift towards using standard models [14]. Research highlights the widespread adoption of the International Classification of Functioning, Disability and Health (ICF) [15], a standard classification developed by the World Health Organization (WHO). Unlike ICD, ICF does not catalog diseases; instead, it describes a person's functioning in terms of body functions, structures (which may be subject to disabilities for various reasons), activity limitations, and participation. Using a comprehensive set of codes taxonomically organized, ICF provides a "snapshot" of a person's functioning at a given moment, capturing both individual and contextual factors. While ICF is less commonly employed in clinical practice than ICD, efforts to make it a full-flagged clinical tool have grown over the years [16]. These efforts include the creation of Core Sets—subsets of ICF codes designed for rapid assessments of functioning in individuals diagnosed with specific conditions.

The taxonomic structure of ICD and ICF, as well as their primary role in healthcare, has led to the development of two important ontologies—BioPortal ICD [17] and BioPortal ICF [18]. Also, local classifications and taxonomies—adopted by national healthcare communities—play a complementary role when paired with ICD and ICF, as they differentiate and expand the perspective of healthcare domains to align more closely with end users' needs. End users may lack deep knowledge of ICD or ICF, making them more comfortable using tools that are familiar to them in their daily practices.

Knowledge and built environment: architecture and BIM. In the built environment sector, ontologies have also been seen as a powerful tool for representing and conveying information about buildings, their components, and their energy performance. In particular, ontologies can be seen as a tool for building information modeling (BIM) [19], a process to develop virtual and multi-dimensional models generated with digital applications. This domain is characterized by considerable complexity, deriving from its multidisciplinary nature (combining architecture, engineering, and information management) and the existence of different standards (local or international, not always reconcilable). This complexity has led to developing and evolving an ontological fragmented scenario characterized by poor interoperability. Among the primary efforts for developing ontologies in this domain, the most discussed in the literature are the following:

- *Industry Foundation Classes (IFC)*: The IFC model was developed as an international standard (ISO 16739:2013) for modeling information related to buildings. Its main purpose is to ensure interoperability between design, construction, and building lifecycle management software. It adopts a geometrically detailed approach to describe built environments, and it is translated into ontology with ifcOWL—but it presents a significantly complex structure, making its implementation difficult. This problem is particularly felt in the scientific community, which has been working for years to try to reduce the complexity of ifcOWL [20, 21].
- *Building Topology Ontology (BOT)*: The BOT model [22] was developed to improve the topology representation and spatial relationships within built environments. It describes spatial and functional relationships between environments, such as rooms, corridors, buildings, and other components, without going into geometric detail as in other BIM models. Topology is represented through entities and properties, such as rooms, floors, and buildings, and defining connections between them (e.g., walls, accesses, doors). However, the geometric representation of spaces is not the model's primary objective. BOT provides a simple and (relatively) limited vocabulary, characterizing itself as a simple and flexible model. This results in greater ease of implementation. Like ifcOWL, BOT is not specifically designed to represent furniture or other functional objects. The lack of geometric definition in the model (in total opposition to the hyper-detailed representation of IFC) allows BOT to focus on environments and their properties while maintaining the possibility of being extended.
- *A Uniform Metadata Schema for Buildings*: Brick [23] was developed to standardize the representation of building data, particularly for building automation and intelligent control. Brick adopts a data-oriented approach, where spaces are represented as "entities" and "subsystems". For these reasons, its use is recommended for specific building automation projects. Additionally, its implementation may require integration with other domain ontologies [24].
- *Building Ontology (BO)*: The Building Ontology model [25] was developed with the goal of facilitating intelligent building management throughout the entire lifecycle, from design to maintenance. Compared to other models (and similarly to ifcOWL), BO is complex to implement as it requires a solid knowledge of the reference ontological model.

Knowledge and aids: a standards-based approach. The possibilities of representing knowledge in the health field have also paved the way for representing aids and assistive technologies. However, the assistive technology sector has only recently been working to identify a common framework [26]. Therefore, the first prototypes of ontologies on this topic do not follow common guidelines. On the contrary, aids are highly regulated in almost all countries and, above all, standardized: the "ISO 9999 Technical Aids for Disabled Persons—Classification" defines a classification and standard terminology for assistive products intended for people with disabilities.

It creates a common language and an organized system for cataloging all those tools, devices, and technologies that help people with disabilities perform daily activities and improve their quality of life.

2.3 The Age-It Ontological Framework

This subsection presents the ontological layer, the "decision-making heart" of the *Age-It* DSS. The ontology engineering process heavily relies on models' reuse, one of the Semantic Web's best practices. Therefore, considering the considerations made in the previous sections, the *Age-It* ontologies were developed following an agile methodology (AgiSCOnt [27]) that allows analyzing the domains in question from different points of view, favoring collaborative knowledge elicitation activities. The *Age-It* ontology is organized as follows: a Common Box imports all the modules that, in fact, make up the ontology. Although not all modules are physically developed as separate files (i.e., the TBox and Abox are modeled within a single.owl file for the smaller modules), the schema depicts all the constituent elements of the ontological layer, also providing the indication of reused ontologies (according to the best practice of the Semantic Web) and the domain to which they are applied. Figure 2 schematizes the modular structure of the *Age-It* ontology.

The ontology editor selected for developing *Age-It* and its modules is Protégé [28], a successful ontology editor now in version 5.5; the editor was selected due to the possibility of modeling rules in "if–then" format using the Semantic Web Rule Language [29], the possibility of using built-in automatic reasoners—for example, Pellet [30]—and, finally, the possibility of testing the ontology with SPARQL query language [31] through the snapSPARQL plug-in [32]. Protégé allows saving in ".owl" files, with various options for syntax serialization: among these, RDF/XML serialization [33], which uses the markup language format to render triples expressed

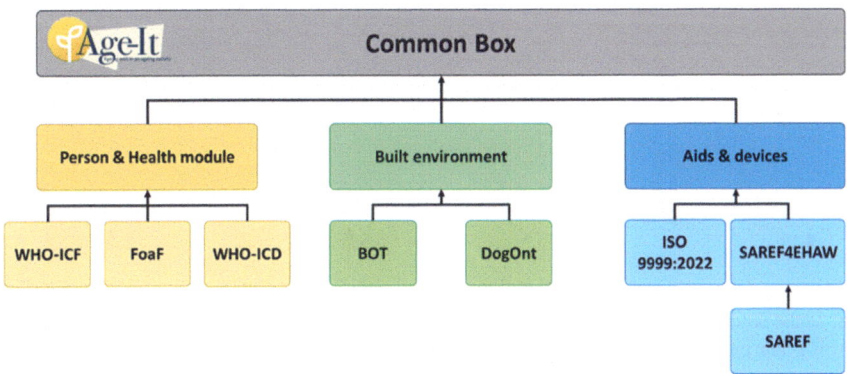

Fig. 2 A schematic representation of the *Age-It* ontology and its "components"

with RDF and OWL, as well as SWRL rules, domains, and ranges of the various properties, and class restrictions.

Person and health condition. In line with what is described in Sect. 2.2, the international classifications ICD and ICF are proposed as ideal candidates for reuse as they are known to healthcare professionals and enjoy broad consensus. The reuse of ICD within *Age-It* is limited to those pathologies that can characterize an elderly inhabitant, while for the ICF, core sets were adopted—following other examples that use ICF core sets as a basis for developing ontological elements (e.g., [34–36]).

From the combination of category codes present in the comprehensive versions of the ICF core sets used to model the personas' conditions, the ICF core that populates the *Age-It* ontology is constituted. From an authoring perspective, this module reuses the ICF ontology already developed and deposited at the OBO Foundry [18]. Adopting a similar approach, the identification of pathologies within the vast ICD classification (version 11) is carried out through a keyword search. This allows us to identify the groups of pathologies that need to be modeled within this *Age-It* module.

To link the occupants to the ICF and ICD codes describing their pathologies and disabilities, an ontology design pattern was reused, establishing an owl:Individual of the class:HealthCondition between the former and the latter; moreover, the more advanced version of the pattern allows detailing the health condition using descriptor individuals (belonging to the class:HCDescriptor) [37]; each descriptor reifies the n-ary relationships between an ICF or ICD code and the related qualifiers [34]. Graphically, the pattern can be rendered as in Fig. 3.

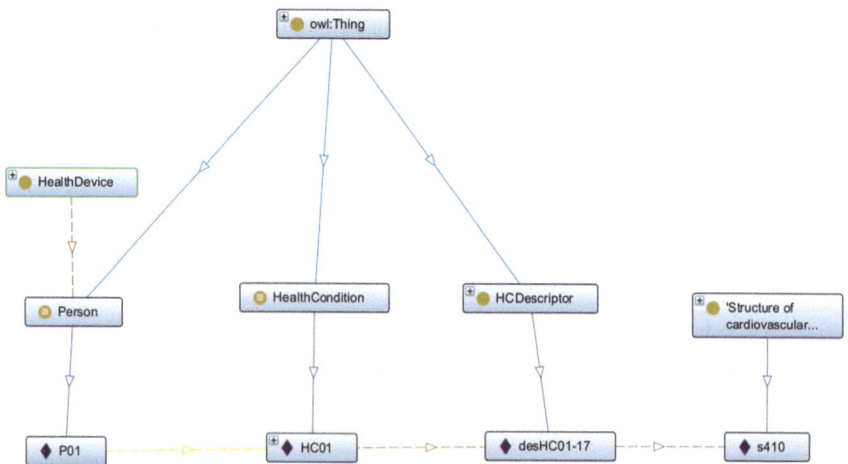

Fig. 3 A (partial) view of the pattern linking a person (P01) with their health condition (HC01) and one of the descriptors describing it (desHC01-17), which involves a specific ICF code (in this case, s4140). The purple prisms represent the owl:Individual, the yellow circles the owl:Class, while the arrows represent different object properties

Reusing the health condition ontology design pattern makes it possible to associate more than one health condition with each inhabitant, to keep track of the evolution of an inhabitant's health condition.

Built environment. The choice of a possible reuse model for the built environment was based on the possibility of adopting a lightweight model (with a low impact on ontological commitment), the possibility of referring to the built environment also in terms of its spaces (e.g., volume, areas, zones, etc.), and the possibility to extend the model to include features of interest for *Age-It*. The lowest level of commitment, while maintaining the ontology's objective, is found in the Building Topology Ontology (BOT), an ontology that reflects the principle of non-overlapping DSS results with what can be developed and implemented using Revit alone.

The built environment module reuses some BOT concepts, among which bot:Element: (a loosely-defined class that includes the physical parts of construction), bot:Zone, and bot:Interface: (a class that includes entities that "interpose" between two elements or two zones, or between an element and a zone). Exploiting the possibility of extending bot:Element, this module represents from scratch:Construction_ element (e.g., doors, windows, and their openings in the walls, pillars, etc.), :Furniture, and:Device. The:Device class is logically equivalent to the homonymous class described in the next subsection. Construction elements are characterized by three subclasses borrowed from the soft reuse of DogOnt [38]. This ontology allows reusing rather common concepts already widely used in other models, such as those related to some construction elements. Additionally, DogOnt was chosen because it contains classes to represent room types and the most common furniture types.

As mentioned in previous paragraphs, BOT does not have a mechanism for describing the geometry of elements or spaces. Remembering that one of the DSS's purposes is to understand the free and occupied space within built environments, the geometry representation uses simple datatype properties—:XDimension, :YDimension, and :ZDimension, which are expressed in cubic meters, and :Weight, expressed in kilograms. With the first three of these minimal properties, it is possible to calculate the value of a fifth property, :BoundingBox, obtained by multiplying the values of the three properties (i.e., calculating the volume of a parallelepiped). This space is obtained to calculate the maximum volume an element can occupy, thus configuring it as an "invisible box" that completely envelops an object or group of objects in a 3D drawing. The calculation of the :BoundingBox datatype property's value is entrusted to an SWRL rule, which uses the language's built-ins to set the parameters and variables for multiplication.

Devices and aids. This module represents the taxonomy reported in the ISO 9999:2022 and leverages SAREF and its extension for eHealth Ageing Well domain [39]. This ontology enables the representation of devices, their services, the power source(s) they adopt, and the health service they deliver. Using saref:Sensor and saref:Actuator classes it also allows for the definition of specific subclasses of saref:Device, which are capable of measuring relevant indoor comfort metrics (e.g., temperature, humidity, air quality, etc.). A particular subclass of saref:Device is s4ehaw:HealthDevice, which identifies those devices created for health purposes; these are then further categorized according to ISO 9999:2022.

Reasoning with *Age-It* (example). The matching between individuals representing users, their health conditions, the devices, the environment and its constraints leverages SWRL rules. This type of condition-action rules enables a deductive reasoning process leveraging arguments represented within the knowledge base. In this way, the original information modelled within the ontology cannot be changed by reasoning results, while new information can be added as a result of a reasoning process. For instance, a person who is in a health condition characterized by J44.1 "Chronic obstruction pulmonary disease with (acute) exacerbations" and associated severe impairments, can benefit from a set of devices, which includes iso9999:2022 04.03.18 "Oxygen units", 04.03.03 "Inhalation equipment", 04.03.12 "Respirators". The matching is represented by the rule:

Person(?p), isInHealthCondition(?p, ?hc), isDescribedBy(?hc, ?desc), ICDCode(?descr1, "J44.1"), ICFCode (?descr2; "b440"), qualifier(?descr2, "3"), 04.03.12 (?resp), suitableFor(?resp, "J44.1"), suitableFor(?resp, "b440"), suggestedForSeverity(?resp, ?x), greaterThanOrEqual(?x, 3) - > suggestedDevices(?p, ?resp).

Therefore, taking into account both the specific disease or impairment and their severity, it is possible to "filter" the best option available. Similarly—and adopting different rules—it is possible to consider some spatial constraints characterizing the inferred devices, to support in the identification of those fitting in the occupant's room.

3 Use Case: An Example of *Age-It* Reasoning

To illustrate the *Age-It* DSS functioning, two use cases are presented (taken from [40, 41]). The first use case refers to a 65-year-old user named Mark, who suffers from a condition that makes him unable to move independently, forcing him to spend most of his day in bed. Mark's caregiver (his wife) needs an aid to help lift him from bed to a wheelchair. The IFC codes describing his condition are d420 Transferring oneself, d450 Walking, d455 Moving around, and d460 Moving around in different locations. Leveraging *Age-It* rules and semantic reasoning, the system retrieves a set of alternative products, associated with ISO category 9999.2022 12.36 "Auxiliary products for lifting people." Among these, within the module "aids and assistive devices" are four types of lifts (Table 1, adapted from [40, 41]).

As illustrated above, each aid is modeled considering its required environmental features. Therefore, by comparing these requirements with the information obtained from the bedroom model represented in the ontology, the system discards the ceiling

Table 1 The devices, aids, and furnishings provided via ontological reasoning for Mark

Devices inferred for mark (Use case 1)			
Persona's health condition	Assistive devices, aids, and furnishings	Aids type/instance id products	Spatial and technological requirements
Reduced mobility (ICF codes d420, d450, d455 d460)	12.36.12 Stationary hoists fixed to walls, floor or ceiling	Ceiling track hoist	– Flat slab – Slab resistance to concentrated vertical loads of 2 kN (200 kg)
		Wall lift (dim. 8 × 115 × 240 cm)	– Max height between floors 350 cm – Absence of false ceiling
	12.36.18 Stationary free-standing hoists	Stationary free-standing hoists (dim. 320 × 420 × 205 cm)	– Free space around the bed of 120 cm
	12.36.03 Mobile hoists for transferring a person in sitting position with sling seats	Mobile hoist (dim. 110 × 110 × 140 cm)	– Free space under the bed of 15 cm – Free passages 110 cm wide – Availability of a circular area with a radius of 150 cm for the movement of hoist

track hoist hypothesis due to the presence of a sloping exposed beam ceiling. Similarly, the hypothesis of a stationary free-standing hoist is discarded because it would require more free space around the bed than the residual volume available.

4 Conclusions and Future Works

The growing need for support among the elderly is driving the widespread adoption of AAL technologies and devices. Care teams must choose these devices based on the individual's specific needs and integrate them into the home environment with minimal visual disruption.

To aid in this decision-making process, the *Age-It DSS* proposed in this study serves as a support tool for designers, ensuring a consistent consideration of the individual's health and functional status, the required devices to enhance autonomy or assist caregivers, and the characteristics of the living space.

The proposed DSS is not meant to replace professionals responsible for recommending aids and devices. Instead, it is designed to be integrated into the decision-making process by leveraging digital information (including individual, device, and space characteristics) to create a virtual exchange platform. This platform enhances collaboration among various professionals, such as healthcare providers and designers, while also supporting the co-design of home environments with end users.

By exploiting Revit's three-dimensional visualization of the environment model, the tool allows for early-stage simulations of AAL device integration, enabling an assessment of their impact on the living space. This approach helps identify the most suitable solutions for the individual's specific needs and guides subsequent design decisions.

Future works foresee developing and testing more use cases to address various living conditions, impairments, disease-related limitations, and aids. In the future, the ontology underlying the system will be populated with several use cases, leveraging clinical collaboration and more matching between aids and the health conditions they address. This would enhance the system's robustness and pave the way for validation with designers and architects.

Competing Interests This study was developed within the project funded by Next Generation EU—"*Age-It*—Ageing well in an ageing society" project (PE0000015), National Recovery and Resilience Plan (NRRP)—PE8—Mission 4, C2, Intervention 1.3".

The authors have no conflicts of interest to declare that are relevant to the content of this chapter.

References

1. Bijak, J., Kupiszewska, D., Kupiszewski, M., & Saczuk, K. (2013). Population ageing, population decline and replacement migration in Europe. In *International Migration and the Future of Populations and Labour in Europe*. https://doi.org/10.1007/978-90-481-8948-9_14
2. Commission, E., Eurostat, Corselli-Nordblad, L., & Strandell, H. (2020). *Ageing Europe—Looking at the lives of older people in the EU—2020 edition*. Publications Office. https://doi.org/10.2785/628105
3. Lloyd-Sherlock, P. (2000). Population ageing in developed and developing regions: Implications for health policy. *Social Science and Medicine, 51,* 887–895. https://doi.org/10.1016/S0277-9536(00)00068-X
4. Tinker, A. (2002). The social implications of an ageing population. *Mechanisms of Ageing and Development, 123,* 729–735. https://doi.org/10.1016/S0047-6374(01)00418-3
5. Ma, C., Guerra-Santin, O., & Mohammadi, M. (2022). Smart home modification design strategies for ageing in place: A systematic review. *Journal of Housing and the Built Environment, 37,* 625–651. https://doi.org/10.1007/s10901-021-09888-z
6. Costa, R., Carneiro, D., Novais, P., Lima, L., Machado, J., Marques, A., & Neves, J.: Ambient assisted living. In *3rd Symposium of Ubiquitous Computing and Ambient Intelligence 2008* (pp. 86–94). Springer Berlin Heidelberg. https://doi.org/10.1007/978-3-540-85867-6_10
7. Rashidi, P., & Mihailidis, A. (2013). A survey on ambient-assisted living tools for older adults. *IEEE Journal of Biomedical and Health Informatics, 17,* 579–590. https://doi.org/10.1109/JBHI.2012.2234129
8. Gruber, T. R. (1993). A translation approach to portable ontology specifications. *Knowledge Acquisition, 5.* https://doi.org/10.1006/knac.1993.1008
9. Spoladore, D., Mahroo, A., Trombetta, A., & Sacco, M.: Comfont: A semantic framework for indoor comfort and energy saving in smart homes. *Electronics (Switzerland), 8.* https://doi.org/10.3390/electronics8121449

10. Spoladore, D., Mahroo, A., Trombetta, A., & Sacco, M. (2022). DOMUS: A domestic ontology managed ubiquitous system. *Journal of Ambient Intelligence and Humanized Computing, 13.* https://doi.org/10.1007/s12652-021-03138-4
11. Graves, M., Constabaris, A., & Brickley, D. (2007). FOAF: Connecting people on the Semantic Web. Cat Classif Q. 43. https://doi.org/10.1300/J104v43n03_10
12. Passant, A., Bojārs, U., Breslin, J. G., & Decker, S. (2010). The SIOC project: Semantically-interlinked online communities, from humans to machines. In *International Workshop on Coordination, Organizations, Institutions, and Norms in Agent Systems.* https://doi.org/10.1007/978-3-642-14962-7_12
13. Katsumi, M. Person ontology. http://ontology.eil.utoronto.ca/icity/Person/1.2/
14. Spoladore, D., Sacco, M., & Trombetta, A. (2023). A review of domain ontologies for disability representation. *Expert Systems with Applications, 228,* Article 120467. https://doi.org/10.1016/j.eswa.2023.120467
15. Üstün, T. B., Chatterji, S., Bickenbach, J., Kostanjsek, N., & Schneider, M. (2003). The international classification of functioning, disability and health: A new tool for understanding disability and health. *Disability and Rehabilitation, 25.* https://doi.org/10.1080/0963828031000137063
16. Stucki, G., Cieza, A., Ewert, T., Kostanjsek, N., Chatterji, S., & Üstün, T. B. (2002). Application of the International Classification of Functioning, Disability and Health (ICF) in clinical practice. *Disability and Rehabilitation, 24,* 281–282. https://doi.org/10.1080/09638280110105222
17. BioPortal: ICD 10 ontology—v. 2021aa. https://bioportal.bioontology.org/ontologies/ICD10
18. BioPortal: ICF ontology—v. 1.0.2. https://bioportal.bioontology.org/ontologies/ICF
19. Demian, P., & Walters, D. (2014). The advantages of information management through building information modelling. *Construction Management and Economics, 32,* 1153–1165. https://doi.org/10.1080/01446193.2013.777754
20. Pauwels, P., Krijnen, T., Terkaj, W., & Beetz, J. (2017). Enhancing the ifcOWL ontology with an alternative representation for geometric data. *Automation in Construction, 80,* 77–94. https://doi.org/10.1016/j.autcon.2017.03.001
21. Wang, C., Zhang, L., & Yan, W. (2024). Enhancement and validation of ifcOWL ontology based on Shapes Constraint Language (SHACL). *Automation in Construction, 160,* Article 105293. https://doi.org/10.1016/j.autcon.2024.105293
22. Rasmussen, M. H., Lefrançois, M., Schneider, G. F., & Pauwels, P. (2020). BOT: The building topology ontology of the W3C linked building data group. *Semant Web, 12,* 143–161. https://doi.org/10.3233/SW-200385
23. Balaji, B., Bhattacharya, A., Fierro, G., Gao, J., Gluck, J., Hong, D., Johansen, A., Koh, J., Ploennigs, J., Agarwal, Y., Berges, M., Culler, D., Gupta, R., Kjærgaard, M. B., Srivastava, M., & Whitehouse, K. (2016). Brick. In *Proceedings of the 3rd ACM International Conference on Systems for Energy-Efficient Built Environments* (pp. 41–50). ACM. https://doi.org/10.1145/2993422.2993577
24. Garrido-Hidalgo, C., Fürst, J., Cheng, B., Roda-Sanchez, L., Olivares, T., & Kovacs, E. (2022). Interlinking the brick schema with building domain ontologies. In *Proceedings of the 20th ACM Conference on Embedded Networked Sensor Systems* (pp. 1026–1030). ACM. https://doi.org/10.1145/3560905.3567761
25. Chávez-Feria, S., & Poveda Villalón, M. Building ontology (BO). https://bimerr.iot.linkeddata.es/def/building/
26. Cook, A. M., Polgar, J. M., & Encarnação, P. (2020). Principles of assistive technology. In *Assistive technologies* (pp. 1–15). Elsevier. https://doi.org/10.1016/B978-0-323-52338-7.00001-9
27. Spoladore, D., Pessot, E., & Trombetta, A. (2023). A novel agile ontology engineering methodology for supporting organizations in collaborative ontology development. *Computers in Industry, 151,* Article 103979. https://doi.org/10.1016/j.compind.2023.103979
28. Musen, M. A. (2015). The protégé project. *AI Matters, 1.* https://doi.org/10.1145/2757001.2757003

29. Horrocks, I., Patel-Schneider, P. F., Boley, H., Tabet, S., Grosof, B., & Dean, M. (2004). SWRL: A semantic web rule language combining OWL and RuleML. *W3C Member Submission, 21*, 1–31.
30. Sirin, E., Parsia, B., Grau, B. C., Kalyanpur, A., & Katz, Y. (2007). Pellet: A practical OWL-DL reasoner. *Journal of Web Semantics, 5*, 51–53. https://doi.org/10.1016/j.websem.2007.03.004
31. Hommeaux, E. P., & Seaborne, A. (2008). SPARQL query language for RDF. W3C Recommendation. http://www.w3.org/TR/rdf-sparql-query/
32. Horridge, M., & Musen, M. (2016). Snap-SPARQL: A java framework for working with SPARQL and OWL. In *Lecture notes in computer science (including subseries Lecture notes in artificial intelligence and Lecture notes in bioinformatics)*. https://doi.org/10.1007/978-3-319-33245-1_16
33. Patel-Schneider, P., & Siméon, J. (2002). The Yin/Yang web. In *Proceedings of the 11th International Conference on World Wide Web* (pp. 443–453). ACM. https://doi.org/10.1145/511446.511504
34. Spoladore, D., Negri, L., Arlati, S., Mahroo, A., Fossati, M., Biffi, E., Davalli, A., Trombetta, A., & Sacco, M. (2024). Towards a knowledge-based decision support system to foster the return to work of wheelchair users. *Computational and Structural Biotechnology Journal, 24*, 374–392. https://doi.org/10.1016/j.csbj.2024.05.013
35. Subirats, L., Ceccaroni, L., Gómez-Pérez, C., Caballero, R., Lopez-Blazquez, R., & Miralles, F. (2013). On semantic, rule-based reasoning in the management of functional rehabilitation processes. In *Management Intelligent Systems: Second International Symposium*.https://doi.org/10.1007/978-3-319-00569-0_7
36. Subirats, L., & Ceccaroni, L. (2011). An ontology for computer-based decision support in rehabilitation. In *Advances in Artificial Intelligence: 10th Mexican International Conference on Artificial Intelligence, MICAI 2011*. https://doi.org/10.1007/978-3-642-25324-9_47
37. Sojic, A., Terkaj, W., Contini, G., & Sacco, M. (2016). Modularising ontology and designing inference patterns to personalise health condition assessment: The case of obesity. *Journal of Biomedical Semantics, 7*. https://doi.org/10.1186/s13326-016-0049-1
38. Bonino, D., & Corno, F. (2008). DogOnt—Ontology modeling for intelligent domotic environments. In *International Semantic Web Conference*. https://doi.org/10.1007/978-3-540-88564-1_51
39. ETSI. (2020). Extension to SAREF—eHealth/Ageing-well Domain. https://www.etsi.org/deliver/etsi_ts/103400_103499/10341008/01.01.01_60/ts_10341008v010101p.pdf
40. Spoladore, D., Romagnoli, F., Ferrante, T., Sacco, M., Mondellini, M., Mahroo, A., & Villani, T. (2024). Customizing seniors' living spaces: A design support system for reconfiguring bedrooms integrating ambient assisted living solutions. In *International Conference on Computers Helping People with Special Needs*.https://doi.org/10.1007/978-3-031-62849-8_46
41. Spoladore, D., Mahroo, A., Sacco, M., Ferrante, T., Romagnoli, F., & Villani, T. (2024). Reconfiguring domestic environments: A decision support system for ambient assisted living. In *2024 IEEE International Conference on Metrology for eXtended Reality, Artificial Intelligence and Neural Engineering (MetroXRAINE)* (pp. 89–94). IEEE. https://doi.org/10.1109/MetroXRAINE62247.2024.10795971

Open Access This chapter is licensed under the terms of the Creative Commons Attribution 4.0 International License (http://creativecommons.org/licenses/by/4.0/), which permits use, sharing, adaptation, distribution and reproduction in any medium or format, as long as you give appropriate credit to the original author(s) and the source, provide a link to the Creative Commons license and indicate if changes were made.

The images or other third party material in this chapter are included in the chapter's Creative Commons license, unless indicated otherwise in a credit line to the material. If material is not included in the chapter's Creative Commons license and your intended use is not permitted by statutory regulation or exceeds the permitted use, you will need to obtain permission directly from the copyright holder.

Social Capital-Rich Places for Healthy Aging

Roberto Di Monaco, Silvia Pilutti, and Marzia Ravazzini

Abstract The architectural design of spaces and enabling technologies should be planned and implemented to facilitate the meeting between people and the development of quality relationships, since these are the real experiences that could change daily life, direct autonomy and facilitate healthy aging. Sociological research and specific policy studies (such as the Italian case study of Trieste) confirm the role of social capital for healthy aging and the possibility of generating it through social mechanisms, such as trust, collaboration, learning, recognition, etc. In this respect, the new territorial welfare structures in Italy—the Community Houses that are currently under construction—could become the hub of this type of social technologies, still largely absent in traditional services dedicated only to technical-health services. Instead, public and third sector social-health workers, together with volunteers and association staff, should promote listening, meeting and collaboration in places open to the community, where they can provide social-health services and live experiences of sociality, altruism, expressiveness, recognition of meanings, search for common solutions to every need. The real challenge is to provide these facilities with spaces and opportunities for physical encounters, inside and around the facilities, available to professionals and citizens. Their vitality will not be long in coming if given the opportunity.

Keywords Social capital · Social mechanism · Healthy aging · Social technologies · Social and health services

---

R. Di Monaco (✉)
Università di Torino, Turin, Italy
e-mail: roberto.dimonaco@unito.it

S. Pilutti
Prospettive Ricerca socio-economica sas, Turin, Italy

M. Ravazzini
Laboratorio Nazionale acceleratore dell'innovazione nelle Case della Comunità, Milan, Italy

1 The Challenge of Community Houses in Italy

This short contribution presents the point of view of the 'National Laboratory, accelerator of innovation in Community Houses', founded in Italy in 2024 by the Association *Prima la comunità*, to stimulate the growth of these new important structures of territorial welfare. The recent reform (DM 77/2022) has in fact modified the organization of territorial welfare in Italy, providing for the construction of 1350 structures in a short period of time (from 2024 to 2026), thanks to the extraordinary NEX Generation Plan key points (PNRR, Mission 6, Health). They should become the most important places within the territorial services system; they include both social and health services, for a catchment area of approximately fifty thousand people each. These structures will therefore be called upon to realize new forms of space open to the public in the community. Inside them, a specific aim is to develop policies for healthy aging, in a socio-demographic context marked by aging and an increased number of people with little autonomy, who live alone without no relatives neither social network. We therefore propose, from a sociological perspective based on available theories and research, just 4 key points that should guide the design of these new places, to effectively associate the architectural solutions and techniques of Ambient Assisted Living (AAL) with what we call the social technologies. From where we stand, the impact of buildings and enabling technologies on the autonomy and the status of health of the elderly would be strongly affected by the way they are designed and socially experienced, and therefore on how people assign meaning to physical and social spaces. These are our 4 key points: (1) the theory of social capital allocates a central role to social relationships that allow people to increase awareness over their own status of health (2) the theory of social mechanisms explains how to create social capital and how social and health workers can strengthen relationships among them and facilitate new relationships both with the people they take care of, and mainly within the community (3) new places as the Community Houses should be designed and experienced to become social spaces where each relationship could grow into collective actions, meaning to increase their capabilities and create the public value (4) the experience of *micro-areas*, a specific social and health structures activated in some neighborhoods of the city of Trieste, in northern Italy, has shown the positive impact on health by the workers embodying these strategies.

2 Quality Relationships for Healthy Aging

Social capital is a key factor for people's health [1], an additional resource to individual skills and resources [2], capable of compensating the negative effects of adverse events on health and influencing the adoption of health-protective behaviors. According to Bourdieu [3], social capital is "the set of current and potential resources that a person can mobilize, linked to the possession of a stable network". It would allow her/him to have access through social relationships to numerous types of resources, for example mutual aid, sharing of information, cooperation for common actions, care of personal safety and property through social control that each person exercises also for the benefit of others, the development of activities and initiatives in the neighborhood, etc. [4]. The most disadvantaged areas usually show less trust and social cohesion. There is a lack of "trust, rules that regulate coexistence, networks of civic associations, or the elements that improve the efficiency of social organization by facilitating the success of initiatives taken by mutual agreement" [5]. Therefore, social capital intrinsically has a dual nature: firstly, it is a social construction, made up of quality individual relationships between people and the symbolic and value-based meanings that are attributed to them [6]. Secondly, it is a resource for people who use it to achieve their goals and maintain autonomy and health [2, 7]. For these reasons, we can measure the existence and quality of social capital by observing interactions, exchanges and cooperation among people, which are bidirectional and multidirectional social relationships that generate effects on behaviors, amplifying intentional behavior and creating empowerment, new skills and a greater degree of autonomy of choice [8]. Recent research has empirically verified the direct influence that social capital has on both individual and collective behavior. Spreading the social norms that accompany stable interactions among people raises it: it happens when people are in the same spaces, look each other into the eyes, talk to each other, do things together, share some part or aspect of their habitat, such as work, living environments, proximity activities and daily life [9].

3 Social Mechanisms and Public Spaces: Places as Opportunities

Social capital is a social infrastructure; it represents the architecture of social relations that characterize groups and sub-groups of people who live and attend any environment. The social capital available to everyone, however, is not only a given characteristic of the social context, but can be created or amplified, as well as reduced or destroyed [8] with the contribution of the same people.

Furthermore, places for public use are important (for example libraries, squares, urban gardens, meeting centers, associations' centers, etc.), where people are physically in a space that exists, it is accessible, available, and where they carry on activities of interest at the presence of others. These spaces offer an opportunity to relate, to dialogue, to collaborate, to live emotions together, to share hopes or projects, to start different experiences which respond to some of their goals, interests or needs. Therefore, social spaces can at the same time develop a new perspective of common action, a construction of endorsed meaning and a joint management of resources by individuals and as a collective contribution.

What is really challenging for healthy aging policies is that these kind of relationships between elderly people and places for public use can represent the lever for both public and private policies and services. The outcome can be reached by activating specific social mechanisms [10]; they are new interdependencies among people, intentionally reproduced by individuals, collective actors or institutional entities, eased by the way in which all spaces are designed. These are social technologies, which require that physical architectures and enabling technologies are planned to enable and provide the same result of physical encounters among people.

Lately, promoting cooperation at the workplace as an organizational strategy has changed the concept of environment: fewer individual offices, more multifunctional spaces, places for meetings and social activities, work groups, connections between spaces in different areas. It has also modified the architecture of technologies: dialogue groups, shared web areas, videoconferencing systems, etc. Similarly, the creation of interdependencies between people and public spaces to share time and activities should change the physiognomy of old clinics and lead to imagining structures designed for relationships and dialogue, both among operators, people accessing services and care and moreover, among these and the professionals themselves. The institutional guidelines for the realization of Community Houses [11] provide many detailed directions to achieve the new structures. However, regarding the social dimension and the functional potential of spaces to generate positive relationships, there would be much to explore and develop. The only 'social' space provided inside or around these large structures is a meeting room, for the use of professionals and sometime the community.

Social and health services are institutional actors that have the peculiar mission of promoting health, especially in disadvantaged areas and with disadvantaged people. According to our evidence, the creation of social capital through their daily way of working, and through the provision of spaces and places open to people's activities, would facilitate the opportunity to increase people's capabilities, especially for the vulnerable ones. This would likely to be, especially to implement individual and collective behaviors oriented towards health.

4 Micro-areas in Trieste: Social Places for Health

The Habitat Microareas (HM) Program is an innovative practice that has shown the effectiveness of social mechanisms and places through which social capital can have an impact on health and reduce health inequalities. Launched in the late 1990s, it has involved active collaboration among public entities, citizens, local associations, social cooperatives and volunteers. The Trieste Health Authority, the Municipality of Trieste and the Public Housing Authority has developed an intervention in the so-called micro-areas, i.e. groups of housing units ranging from 500 to 2000 inhabitants each, with high social-health fragility. A multidisciplinary team of health professionals and social workers worked daily in each micro-area to identify and respond to the community needs, facilitating access to services and promoting mutual aid initiatives. Evaluative research has been conducted on the experience, which highlighted the social technologies adopted and the use of places [12].

First of all, the service headquarters was located right in the center of the micro-area, in a public housing apartment, becoming the operational center for social and health services, but at the same time it has been a community space attended by citizens to spend time in, meet others, realize daily life common activities, such as cooking, eating, etc. and also expressive and cultural activities, such as discussions, training, plays, etc. The headquarters therefore became a center for meeting and a place to build relationships and do common activities, which stimulated the active participation of citizens. The place was architecturally poor, but still well recognized by people as their own, always accessible, inserted in a full-continuity to their living environment, rich in functionalities, really significant for people. Exploring on people's favorite meeting places, the community spaces within the Micro-areas emerged central to social interactions, overcoming bars and public gardens. Secondly, social and health workers considered the building, the neighbors, those who lived near the house, as part of the intervention. Therefore, the workers contributed to introducing people and to growing constructive relationships: going shopping for someone who cannot go out, helping a neighbor in a moment of difficulty, could start new relationships, trigger reciprocity, change the way of perceiving the social context. We remark the role of the worker as a promoter of trust, made possible by the fact of being a habitual presence in the field of the micro-area and, therefore, in the social context where people do live. Finally, the professionals have aimed to facilitate the person's paths and relationships towards the service system and bureaucracy, to overcome critical issues at a social, economic and environmental level. In this case, professionals moved as if their task was to grow an active network around the person, so that they could easily relate to distant, inaccessible services, and resolve critical issues, etc.

By studying the impact of this service, crucial changes in the behavior of residents were observed, with a significant increase in cooperation and social commitment. People proved to be more active and willing to help each other and to use common spaces for shared activities. Teamwork among professionals led to improvements in the services offered, and in personalization. Furthermore, the capabilities to manage adverse events improved, exercising freedom of choice, with increased quality and variety of the kind of help received, as well as a greater degree of satisfaction with the assistance obtained.

The outcomes indicate that open and functional spaces for meeting people and attention to the context of social relations can create more cohesive and resilient communities, capable of greater self-management and collective response to social and health needs. This experience demonstrates that the virtuous combination of spaces, people and social technologies can lead to substantial improvements in the management of public health and community well-being.

References

1. Moore, S., Stewart, S., & Teixeira, A. J. (2014). Decomposing social capital inequalities in health. *Epidemiology and Community Health, 68*, 233–238.
2. Coleman, J. S. (1990). *Foundations of social theory*. Harvard University Press.
3. Bourdieu, P. (1986). The forms of capital. In J. G. Richardson (Ed.), *The handbook of theory: Research for the sociology of education* (pp. 241s–2258). Greenwood Press.
4. Carpiano, R. (2008). Actual or potential neighborhood resources for health. In I. Kawachi, S. V. Subramanian, & D. Kim (Eds.), *Social capital and health* (pp. 83–93). Springer.
5. Putnam, R. (1993). *La tradizione civica nelle Regioni italiane*. Leonardi Editore.
6. Kawachi, I., Subramanian, S. V., & Kim, D. (Eds.) (2008). *Social capital and health* (pp. 1–26). Springer.
7. Moore, S., & Kawachi, I. (2017). Twenty years of social capital and health research: A glossary. *Journal of Epidemiology & Community Health, 71*, 513–517.
8. Di Monaco, R., & Pilutti, S. (2021). *Scommettere sulle persone. Leadership distribuita per l'organizzazione smart & green, agile, lean e 4.0* (2nd ed.). Egea Bocconi.
9. Centola, D. (2018). *How behavior spreads: The science of complex contagions*. Princeton University Press.
10. Barbera, F. (2004). *Meccanismi sociali*. Il Mulino.
11. Mantoan, D., di Lavoro, G., Borghini, A., Fortino, A., Riano, F., Izzi, A., Capolongo, S., Buffoli, M., Gola, M., Brambilla, A., & Mangili, S. (Eds.). (2022). Documento di indirizzo metaprogetto Casa di Comunità. In *Quaderno di Monitor 2022*, Supplemento a Monitor, AGENAS, Agosto.
12. Di Monaco, R., Pilutti, S., d'Errico, A., & Costa, G. (2020). Promoting health equity through social capital in deprived communities: A natural policy experiment in Trieste, Italy. *Social Science & Medicine—Population Health, Elsevier, 12*, 100677 1–11.

Open Access This chapter is licensed under the terms of the Creative Commons Attribution 4.0 International License (http://creativecommons.org/licenses/by/4.0/), which permits use, sharing, adaptation, distribution and reproduction in any medium or format, as long as you give appropriate credit to the original author(s) and the source, provide a link to the Creative Commons license and indicate if changes were made.

The images or other third party material in this chapter are included in the chapter's Creative Commons license, unless indicated otherwise in a credit line to the material. If material is not included in the chapter's Creative Commons license and your intended use is not permitted by statutory regulation or exceeds the permitted use, you will need to obtain permission directly from the copyright holder.

The Right to Grow Old in Your Own Home

Claudio Falasca

Abstract Frailty due to age puts into question the right to grow old in one's own home. Between institutionalization and isolation, it is necessary to identify a third way. A new housing model that conceives living as an integrated service. This idea is outlined in the Delegated Law for the Reform of Elderly Care. The assisted housing model is the one that best responds to this need.

Keywords Living · Elderly · Frailty · Assisted housing

1 Introduction

There appears to be little need to further elaborate on the right to grow old in one's own home: the entire research program on the "habitable future", addressed in the present edited manuscript that includes this contribution, is aimed at this goal.

The problem is that today this right is not guaranteed. The pandemic has confronted us with this harsh reality. It is no longer enough to own your own home (about 80% of elderly people live in homes they own) to be sure of being able to enjoy this right if the home and its context are not "equipped" for the new needs of an increasingly long-lived population.

As AUSER,[1] and as *Abitare&Anziani*,[2] we have repeatedly called for greater attention to this need: in fact the problem has always been evaded.

---

[1] https://www.auser.it/.
[2] https://www.abitareeanziani.it/rivista/.

C. Falasca (✉)
AUSERNazionale Studies Office, Rome, Italy
e-mail: c.falasca@auser.it

Abitare e Anziani, Rome, Italy

The paradoxical thing is that we are talking about a problem that affects millions of families... and yet it struggles to become a major social issue. It seemed that the pandemic had at least the merit of making us understand this. Unfortunately, however, once the critical phase was over, everything stopped.

The pandemic. The impact of the pandemic on frail people has highlighted the limits of the current housing model in its fundamental dimensions: the home environment, the building, the neighbourhood [1]. Living environments designed and created by taking as a reference the stereotype of the young, male person, at the height of his abilities, in the face of a society characterized by the progressive growth of lonely and frail elderly people. A growing number of people who, aware of the risks of frailty and even if they own the home in which they live, in order not to find themselves trapped in the grip of "institutionalization—care—isolation", rightly want to choose where to live and with whom to share the accommodation. They express the request for an environment suited to their needs that guarantees social inclusion, protection and independent living: all things that the current housing model does not ensure.

2 The PNRR, Building Bonus, Delegated Law 33/2023

The National Recovery and Resilience Plan [2], (PNRR) designed to remedy the damage caused by the pandemic, dedicates resources to these non-secondary needs. The goal is to implement interventions in favor of vulnerable people by intervening both on their housing conditions and on local territorial services.

Unfortunately, at present, there are all the elements to think that a unique opportunity is being missed as there are no visible signs of a change of direction.

Going to verify the content of the PNRR intervention lines on Urban regeneration and social housing, what emerges is a traditional approach completely unaware of the requests coming in particular from the WHO regarding ageing in place [3], the age friendly city [4], ambient assisted living (AAL) [5], universal design [6].

Perhaps it was thought that, since elderly people are largely owners of the houses in which they live, by making adequate incentives available to them (the building Bouss) they would decide autonomously how to improve their living conditions. Unfortunately, we know how it ended: the building bonuses, as conceived, have committed enormous public resources to interventions that perhaps, if better directed, could have solved many problems. For example, by favoring the removal of barriers, starting with the construction of elevators in public and private homes, rather than the renovation of facades.

But as mentioned, the needs of fragile people cannot be solved by intervening only on the physical and technological characteristics of homes, buildings and neighborhoods, but it is also necessary to intervene on the quality of the system of local, civil and social services.

Also on this front, the PNRR has provided for the reform of assistance for the elderly. The drafters of Delegated Law no. 33/2023 [7], did not miss the importance of the issue when the text (Arts. 2 and 3) indicates the need to intervene on the quality of residential services, community services and housing conditions. And even if it is not specified exactly in what sense the requested 'qualification' should take place, the Delegation nevertheless expresses the need to develop new housing models that take greater account of the needs of vulnerable people.

Furthermore, it is expected that the measures aimed at improving the quality of housing from the point of view of the functional characteristics of homes, buildings and neighborhoods must be integrated with those aimed at supporting the needs of social inclusion, protection and independent living.

The serious thing is that a year after the approval of Legislative Decree 29/24 [8], implementing the Delegation, nothing has yet been done to give substance to the planned measures and, therefore, despite having the means and opportunities and models to refer to, it has not been possible to develop and experiment with a model of progressive integration between the hard and soft dimensions of housing quality.

3 Assisted Housing

According to both national and international experiences, in my opinion the optimal reference model is that of "assisted housing". This means an apartment, more often small in size, which generally accommodates a family of elderly people or a single elderly person who maintains a certain degree of autonomy and therefore is able to self-manage the main activities of daily life autonomously, supported by technologies and personalized services provided at their home.

This type of housing, called in different ways: protected housing, social housing, mini protected housing, mini housing for self-sufficient people, apartment groups etc., is united by the same approach of allowing those who are advanced in years to live autonomously, leveraging the ability to continue to maintain their own rhythms of life in an adequate context, thanks to external support.

It is in this context that the results of the valuable research work on spaces, objects and devices to support aging carried out by the "habitable future" program precipitate in an optimal and concrete form.

In short, assisted housing is able to guarantee an integrated housing service of: monitoring and safety of life at home; support for relationships; support for autonomy; support for social inclusion.

4 Conclusions

The implications of a housing policy aimed at fragile people centered on the idea of "assisted housing" are not overlooked:

- with them the idea of "home" evolves towards the "integrated service" that is not limited to the ownership of the home, but is enriched by a plurality of housing and proximity services (civil and social-health) to be used individually and collectively;
- the WHO indications regarding ageing in place, age friendly cities by the WHO will have to be translated into binding references by reviewing the now obsolete legislation on building and urban standards;
- the evaluation of housing quality should not be limited to compliance with mere "quantitative standards", but should include the adoption of integrated evaluation models on the enabling or disabling capacity of living environments, already in use in other countries (the English Housing Quality Indicators Form, or the American Livability Index);
- as the service function becomes more important, the owners of welfare policies take on a greater role. In other words, the time has come to require strong integration between housing policies and social and health policies at all institutional levels;
- it will also be necessary to involve (train) new skills, in particular on the social intervention front, engaging them in particular in the management of public residential building complexes in order to overcome the merely "real estate-administrative" management of ATER;
- particular attention should be paid to policies to support supply, in particular public supply, but also in the public/private relationship;
- certainly the "management" of this new housing model will require a new "culture of living" and the commitment of new skills (training) on the front of social and socio-health intervention in proximity services, assistance in the use of new technologies, etc.;
- the complexity of this housing model cannot be grasped in its entirety by relying only on the public role and the market. Instead, it will be essential to involve Third Sector organizations in co-programming and co-design initiatives within territorial programs of housing policies and urban regeneration. It is only at this level that it will be possible to fully grasp the social demand in all its articulation and specificity.

To do all this, public and private resources will undoubtedly be necessary, but above all it will be essential, unlike what has not been done with the "bonuses", or with the PNRR, a public governance capable of intelligently directing the necessary individual and collective actions.

References

1. Italian Government. (2021). *National recovery and resilience plan.* Retrieved March 5, 2025, from https://www.governo.it/sites/governo.it/files/PNRR_0.pdf
2. World Health Organization. (2007). *Global age-friendly cities: A guide. World Health Organization.* Retrieved March 5, 2025, from https://www.who.int/publications/i/item/9789241547307
3. European Parliament, Council of the European Union. (2008). *Decision No 742/2008/EC on the Community's participation in a research and development programme undertaken by several Member States aimed at enhancing the quality of life of older people through the use of new information and communication technologies.* Retrieved March 5, 2025, from https://eur-lex.europa.eu/legal-content/EN/TXT/?uri=celex%3A32008D0742
4. The Center for Universal Design. (1997). *The principles of universal design,* Version 2.0. North Carolina State University.
5. Forum Terzo Settore. (2023). *Il Pnrr, le politiche sociali e il Terzo settore.* Retrieved March 5, 2025, from https://pnrr.forumterzosettore.it/wp-content/uploads/2023/07/report_forum_terzo_settore_7.pdf
6. Italian Law 23 march 2023, No. 33. Government mandate on policies supporting older people.
7. Falasca, C. (2023). Ripensare le politiche abitative, *Abitare e Anziani,* 3, pp. 5–8. Retrieved March 5, 2025, from https://www.abitareeanziani.it/wp-content/uploads/2023/09/AeA_3_2023.pdf
8. https://www.gazzettaufficiale.it/atto/serie_generale/caricaDettaglioAtto/originario?atto.dataPubblicazioneGazzetta=2024-03-18&atto.codiceRedazionale=24G00050.

Open Access This chapter is licensed under the terms of the Creative Commons Attribution 4.0 International License (http://creativecommons.org/licenses/by/4.0/), which permits use, sharing, adaptation, distribution and reproduction in any medium or format, as long as you give appropriate credit to the original author(s) and the source, provide a link to the Creative Commons license and indicate if changes were made.

The images or other third party material in this chapter are included in the chapter's Creative Commons license, unless indicated otherwise in a credit line to the material. If material is not included in the chapter's Creative Commons license and your intended use is not permitted by statutory regulation or exceeds the permitted use, you will need to obtain permission directly from the copyright holder.

Waiting Spaces in Primary Care for Healthy Ageing: Applying Design Guide-lines to the Project

Elena Bellini and Nicoletta Setola

Abstract Local socio-healthcare facilities for primary care are drivers for health promotion in the social community. This three-year research project, led in "Age-It", deals with the study of waiting spaces, focusing on sensory environments and integrated technologies to support physical and mental wellbeing and to promote healthy and active ageing in the Casa della Comunità (House of the Community), which is the Italian new socio-healthcare facility. The Design Guidelines are one of the main results of the research. This paper presents an example of the application of this result by using Design Guidelines to develop the architectural project and the integration of advanced technologies in the waiting spaces of Casa della Comunità Le Piagge in Florence. After an introduction about the research process, the paper describes the structure of the Design Guidelines, composed by design strategies and sheets. Then, the architectural project is presented according to the main strategies, dealing with: create easily accessible places and paths; create environments for the person's psychological comfort and support; involve, welcome and encourage sociality; promote active waiting time; promote and activate health. The original contribution to promote healthy ageing is in transforming waiting into active waiting time, doing different activities which support mental and physical wellbeing, prevent chronic diseases and contribute to diffuse knowledge on healthy lifestyles.

Keywords Primary care facilities · Health promotion · Active ageing · Design for health · Sensory design

---

E. Bellini · N. Setola (✉)
Department of Architecture, TESIS Research Center, University of Florence, Florence, Italy
e-mail: nicoletta.setola@unifi.it

1 Introduction

1.1 The Research Topic

The research deals with the study of waiting spaces of one of the socio-health care facilities for primary care in Italy that is the House of the Community (CdC), to promote healthy and active aging and support physical and mental wellbeing through sensory environments and integrated technologies.

It is now consolidated in the scientific literature that the built environment is one of the determinants of health as it acts as a promoter of healthy lifestyles: doing physical activity, having a healthy diet, being involved in positive social interactions. Healthy lifestyles help to prevent chronic diseases (such as cardiovascular diseases, diabetes, some types of cancer) and physical and cognitive decline. Thus, actions aimed at supporting fragility and combating loneliness also benefit from the design of a healthy and inclusive built environment.

The research aims at:

- reducing patients/caregivers' stress during the waiting time, before the healthcare service;
- increasing people's psychological comfort and support;
- favouring health promotion: healthy and active aging and prevention of chronic diseases.

These main objectives refer to three fundamental theories:

- the influence of the built environment on people's health by altering stress levels, referring to four specific dimensions at the building scale: stimulation, coherence, affordance and control [1];
- coping [2] and restoration [3–6], i.e. those characters of the environment that allow patient and visitor to recover and regenerate;
- health-promoting architecture [7] and buildings [8], interpreting the concept of health promotion [9] through the built environment, i.e. active design, promotion of physical activity, healthy buildings [10], etc.

For this reason, starting from the national Requirement Classes of the Building System (UNI 8289:1981), this study focuses on the Wellbeing and Usability classes, also according to the Guidelines for the Humanization of Healthcare Spaces [11]. According to these principles, the research deals with:

- Psycho-emotional well-being and the needs of: privacy, concentration, social interaction, continuity with the home environment, mental disengagement (positive distraction and restoration), psychological support, control, information and involvement;
- Environmental well-being, i.e. the conditions that guarantee physical and sensory well-being: acoustic, thermo-hygrometric, visual, olfactory, and tactile well-being;

- Usability, i.e. the set of conditions that allow the use of spaces, furniture and equipment in adequate conditions. More specifically, we analyze the layout and configuration of spaces.

1.2 Age-It Research for Healthcare Waiting Spaces

This research is led in "Age-It—Ageing well in an ageing society" project.

During the three years of the project many research actions were carried out to develop Design Guidelines for waiting spaces in primary care facilities to promote active and healthy ageing. The research process will be briefly described in the next paragraphs (Fig. 1). The details of each phase of the research and used methods will be presented in a specific publication.

Background research. First of all the literature review in the field was investigated. From the analysis of the literature, a lack of guidelines for the design of welcome spaces in primary care facilities was identified. For this reason, an analysis of best practices was developed to fill the gaps in the literature through the application of significant technical and design solutions in the specific context of waiting spaces in primary care facilities: the identification of invariants in all case studies, which can be translated into input for the project; as well as the investigation of new concepts for innovation coming from other healthcare contexts or related to different users, to be reinterpreted and applied to the CdC project.

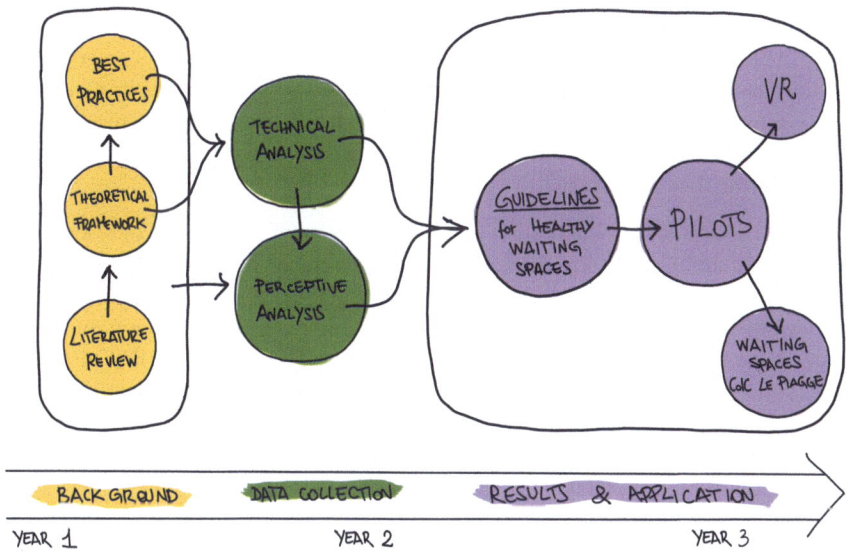

Fig. 1 The research process

A Theoretical Framework was defined for the selection and analysis of case studies, according to three main areas of investigation: Welcome and Waiting spaces; Primary socio-healthcare facilities; Sensory Design.

Data collection and field research. Using the background research as a basis, two types of analysis were performed: a technical one, developed by the researchers of the architectural field, who were collecting data from the technical analysis of international best practices, and a perceptive analysis, made in collaboration with IRCCS INRCA (Task 1.5, WP 1, spoke 9 of Age-it team project). The results of the technical analysis therefore were the basis of the perceptive analysis, submitting interviews and questionnaires which involved directly the users of the waiting spaces (patients, caregivers and healthcare professionals).

Technical analysis. Welcome and waiting spaces of various types of health and socio-healthcare facilities (e.g. hospitals, clinics, health centers, etc.) were analysed. European buildings were selected in order to evaluate examples close in size and in the social and cultural context to researchers' case of application, and only buildings designed and built since the 2000s. Specialized healthcare facilities were excluded, such as emergency departments, cancer centers, pediatric hospitals, dental clinics, etc. as they would have different requirements and contexts. Looking for innovation, Sensory Design has been identified as an adequate approach to solve both the objectives of psychological comfort and health promotion, to provide relaxing, welcoming and customizable environments for different types of users through integration of digital technologies. For this reason, an analysis of best practices in the field of sensory design has been carried out to understand how to apply these solutions to the specific context of the research.

Sensory solutions integrated in European healthcare facilities, designed and built since 2010, were analysed. Assisted residential facilities and day centers were excluded as they differ in the types of requirements, spaces and use, and solutions related to waiting areas were favoured. In total, 37 case studies were analysed, producing 37 design sheets as a synthesis of the technical analysis, representing one of the intermediate results [12].

Perceptive analysis. In collaboration with IRCCS INRCA, an Institute of research in the Geriatric and Gerontological fields, a semi-structured interview was submitted to healthcare professionals (doctors, nurses, etc.), patients and caregivers of relevant healthcare facilities in order to collect data on the use and perception of the spaces expected to promote healthy and active aging. The context is that of outpatient clinics and House of the Community, in order to use results as part of the tools for developing guidelines. The interviews were submitted in INRCA facilities in Ancona (Health Director, Department of Urology, Department of Gastroenterology, Antidiabetic Center), Casa della Comunità Le Piagge (Florence), Casa della Comunità Morgagni (Florence), Department of Multidimensional Medicine of Florence. These results are qualitative and have been integrated in the Design Guidelines development. After the interviews, a questionnaire has also been prepared to be submitted to patients and caregivers, to provide more quantitative data in the same facilities where the interviews were carried out, as results to be integrated in the Design Guidelines.

Results and application. Design Guidelines for waiting spaces in primary care facilities to promote active and healthy ageing represent one of the main results of the research and are described in Sect. 2. The Design Guidelines have been applied in two pilots to be verified and integrated in the next year of research: a digital pilot (using virtual reality) and a physical one. In the last section of this paper the physical pilot will be described as an example of a project of renovation of waiting spaces in a CdC in Florence to promote active and healthy ageing (see Sect. 3).

2 Design Guidelines for Waiting Spaces in Primary Care to Promote Active and Healthy Ageing

2.1 Design Guidelines Structure

The Design Guidelines are structured according to "general strategies" to promote health through the built environment, splitted down into more specific "actions" and "sub-actions" for the project, referring to different areas of intervention, and "design sheets" which identify the space requirements to promote healthy and active ageing and some specific solutions for the integration of digital technologies in the health promoting waiting spaces.

The objective of the "design strategies" is to indicate the actions necessary to promote healthy and conscious lifestyles and behaviours through the built environment and to design waiting spaces, able to improve people's health and well-being conditions and, more specifically, healthy and active aging. These strategies represent support in decision-making processes, such as briefing, programming and planning of refurbishment or new construction interventions in waiting spaces. Strategies provide an indication of which actions are necessary when developing the project and therefore what needs to be foreseen to promote health through spaces.

The objective of "design sheets" is to provide design requirements for waiting spaces and the integration of digital devices, like as an operational tool of concrete support in the design phase of the refurbishment or new construction of the waiting spaces. The design sheets provide indications on which requirements, individual solutions, interventions, and measures, even on a detailed scale, are necessary during the development of the project of health promoting waiting spaces and the subsequent steps.

2.2 Design Strategies

The strategies that have been considered useful in promoting healthy and active aging through waiting spaces are the following.

Create easily accessible places and paths. It is necessary to configure the waiting spaces in layouts that optimize the use and the quality of healthcare environments and favour orientation and wayfinding by clarity and coherence of spatial configuration, as well as visibility, permeability, accessibility, and usability of spaces and objects.

Involve, welcome and encourage socialization. The CdC is a place that first of all should make people feel included by the creation of intergenerational and multicultural spaces for all, as well as developing a functional mixité, and promoting solutions that foster relationships and sense of community, involving citizens and organisations of the third sector. It is also essential to encourage a sense of familiarity, through the use of colors and materials that reduce the institutional aspect of the environment, the placement of different types of furniture for all, configured in layouts which encourage relations between users, and the integration of technological devices to promote interactions with the space.

Create environments for the person's psychological comfort and support. Positive distraction and restoration in waiting spaces promote psychological comfort, stress (and consequently aggressions) reduction and a general positive healthcare experience. Restoration is provided by the relationship with nature (indoors or outside, both directly and indirectly); the use of illustrations, colors and artworks, also involving citizens for local arts; the interaction with the space or technological devices to play and relax. Comfort and physical well-being are firstly favored by the quality of spaces and the indoor environmental comfort: use of natural light; luminous quality, including artificial lights; natural ventilation; acoustic comfort, etc. Customised solutions that allow direct control over the comfort state (temperature, brightness, etc.) by users are preferable. Automation technologies can support this process, as well as promoting sensory well-being and regulation. Privacy also contributes to comfort, especially facing the outpatient area and medical rooms, where patients and caregivers should feel at ease when talking with healthcare professionals.

Promote active waiting time. The role of waiting spaces and interactive technologies is to promote an active waiting time, in which the patient or caregiver does not just wait, but uses this time to promote healthy and active ageing. This can happen in several ways, including: improve knowledge on health issues and on the assistance and support services provided for in the CdC; relax and positive distract through immersive sensory scenarios; do cognitive training, also having the possibility of self-testing; do physical activity, both encouraged through the interaction with the space and digital technologies, and through passive training devices, as well as providing the possibility of self-assessment to prevent chronic mental and physical diseases.

Promote and activate health. The design of buildings that promote health through the built environment, i.e. health promoting buildings and active design solutions represent a basic strategy and actions for spaces to promote healthy and active ageing. Promoting health means also the dissemination of knowledge on health issues and on the assistance and support services provided for in the CdC and, more generally, healthy lifestyles, access to healthy food, drinking water, breastfeeding area, promotion of the use of stairs, etc. This strategy also includes actions to promote movement and gentle physical activity within the waiting spaces and in the contiguous external spaces, digital interaction for users' self-assessment and cognitive training.

3 The Pilot: Casa della Comunità Le Piagge Project

3.1 CdC Le Piagge: The Context

Primary care services in Italy are evolving in response to the health reform DM 77/2022, which is promoting the spread of socio-healthcare facilities (e.g. House of the Community) on Italian territory to support well-being and health promotion in communities. The House of the Community (CdC) is a new model of primary socio-healthcare facility with the aim of treatment, prevention and health promotion, focusing on chronic diseases and promoting continuous care for the elderly. It is a physical place within the healthy neighborhood that can facilitate health and social assistance, and offers access to information. For this reason, when entering into a CdC, people should be welcomed into a comfortable and safe space, especially in cases of emotional and social vulnerability, where professionals can assist people's social and health needs and promote their well-being and inclusion in a healthy and equitable community. Le Piagge is one of the CdC of Florence city in Tuscany Region currently undergoing redevelopment (Fig. 2). The facility is a courtyard building housing the outpatient rooms in the front and the rear wing. The layout configuration of each wing is characterised by a fairly wide central corridor, onto which the outpatient rooms face. The areas used as research pilot are the corridors in the rear wing, the atrium, two waiting rooms and an outdoor terrace.

3.2 Concept of the Project

The project is founded on the application of Design Guidelines and was developed starting from the general strategies for waiting space to promote healthy and active ageing. The main idea is to transform waiting into active waiting time by doing different activities which promote health (Fig. 3). Health is considered both as a physical and mental wellbeing.

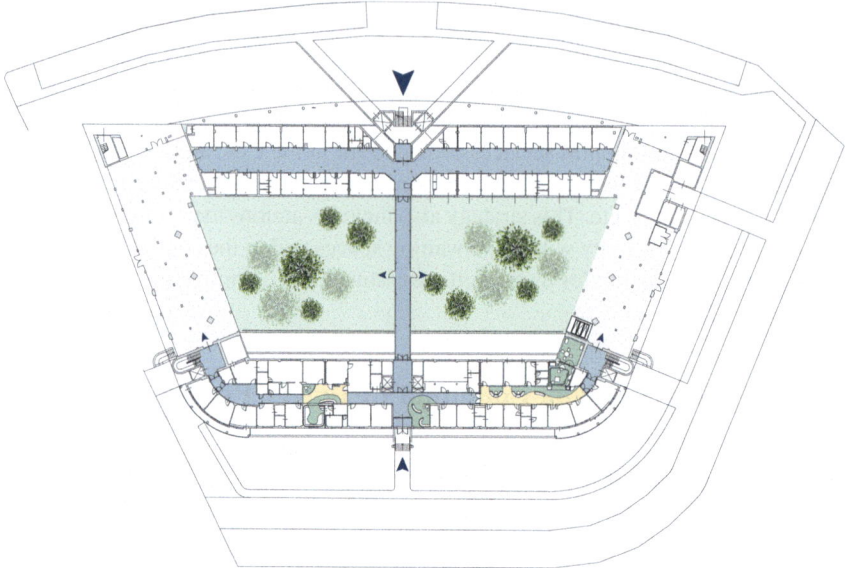

Fig. 2 General plan of the CdC Le Piagge

Promoting health for the body during waiting time means informing people on healthy lifestyles and behaviours, promoting diffuse physical activity by the design of the space and the integration of technologies and special furniture: favouring active moving such as gentle short-term physical exercise; permit passive physical activity such as sitting on special furniture that make people active while sitting; promote auto-assessment of physical condition to prevent chronic diseases.

Thus, promoting health for the mind means let people feel restored by art or nature; socialising and making relations with other patients or caregivers; having comfort during waiting time to improve psychological support; positively distract before the medical services to relax and reduce stress, especially for people with frailties; doing cognitive training to exercise the mind potentials; and doing auto-assessments to prevent mental diseases, especially related to ageing, such as dementia.

3.3 Project Description

The new areas of waiting in CdC Le Piagge were developed to answer to different needs of the different users of the CdC: people of all ages and cultures (in this area there are many people coming from differents countries, such as a big Chinese community) going to the General Practitioners and specialist medical services, referring particular attention to older people and people with chronic diseases; children and adolescents going to the psychiatric and rehabilitation services, accompanied

Fig. 3 Concept of the project: health promotion

by their parents (often the mom with other sons or daughters), referring particular attention to people with cognitive and neurodevelopment diseases; people coming for blood sampling or other medical exams; people looking for social assistance; people going to the continuity of care or rapid intervention points (PIR service).

To address the multiple needs of the many common users of the CdC, the researchers were designing different areas (Fig. 4, area A and B; Fig. 5, area C) with different functions, spread in the connections and waiting spaces of the south wing of the CdC Le Piagge. The idea of a diffuse waiting space makes it possible to find and choose the best place to wait according to people's preferences.

Below are the main concepts of the design project will be described, according to the strategies developed in the Design Guidelines.

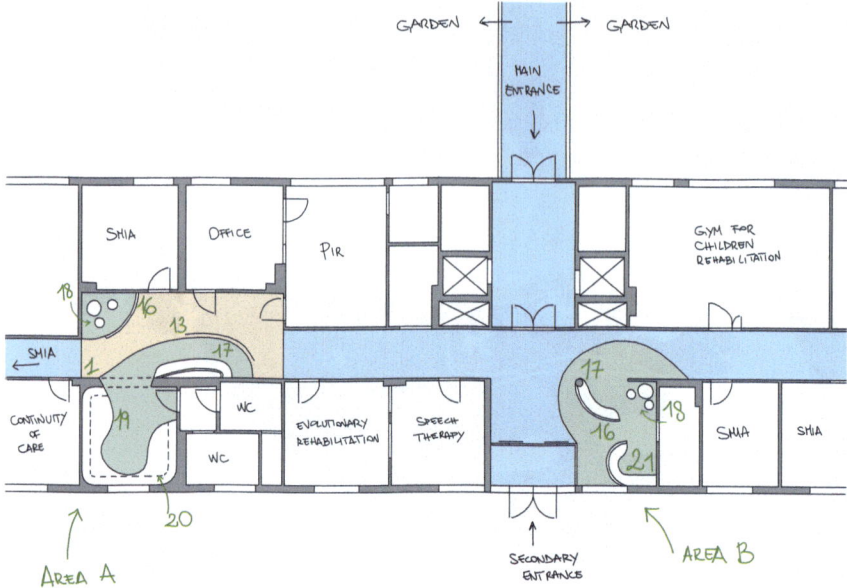

Fig. 4 Plan of area A and B: 1—portal; 13—curved line track of lights; 16—timber screen; 17—curved line sofa for short stay; 18—children playing area; 19—composition of a curved line sofa with a soft wall to sit or lay down; 20—upper part of the soft wall with integrated RGBW light to enlighten the ceiling and warm soft light to enlighten the sofa, audio diffusers and a video projection on the ceiling; 21—breastfeeding area, with integrated audio diffusers and RGBW lights

Create easily accessible places and paths. The CdC has a single entrance, so the connections and the wayfinding system is quite clear and comprehensible. Anyway, to improve the clarity of paths, connections are highlighted by colors of the flooring: the existing area has a light blue colour (Fig. 4) while the new areas are differentiated by the use of orange and green flooring (Figs. 4 and 5); the orange colour represents the way out (also as direction for the fire safety) and the green highlights the place to stay and wait. The green colour also creates a link with nature and it is a colour which usually makes people feel calm and relaxed. All the waiting areas are connected by the main central corridor, so from every part of the CdC the vision of the green colour immediately gives the comprehension and identification of spaces where to go and to stay while waiting.

The visibility and permeability of spaces are always guaranteed and the connection between the interior spaces and the exterior area is favoured in every waiting area. In some areas, such as the corridor, specific waiting spaces are filtered by timber semi-transparent screens (16). These permeable walls aim at isolating areas that need more privacy, such as the breastfeeding area (21), the children's playing areas (16), the cognitive training "station" (a), the physical activity "station" (b), etc., but maintaining the continuous visibility of spaces to guarantee patients' and caregivers' safety. Moreover, the continuity of care and PIR service (Fig. 4) are always visible

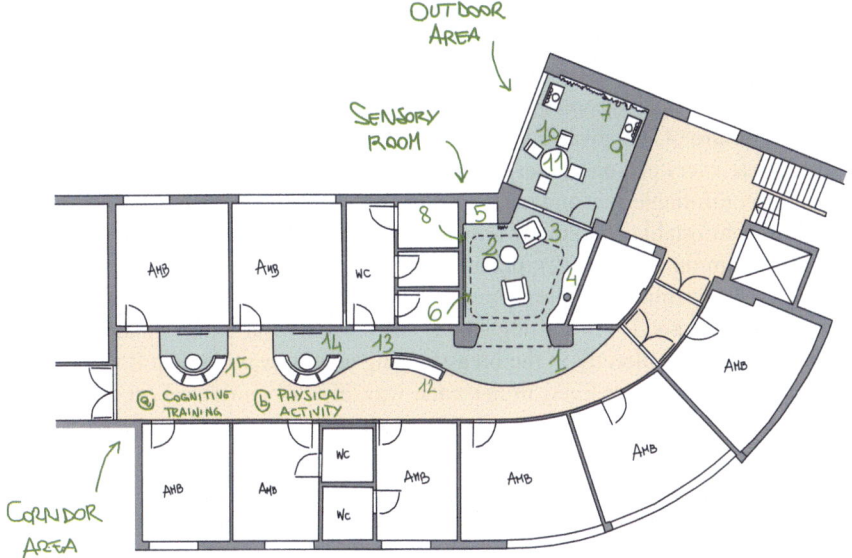

Fig. 5 Plan of area C: 1—portal; 2—poufs; 3—armchair; 4—composition of a curved line sofa and a soft wall to sit and be contained; 5—sensory pod with integrated RGBW lights; 6—curved line false ceiling to integrate RGBW lights; 7—green wall; 8—video projection on the wall with integrated audio diffusers; 9—H bench alpha; 10—active bench; 11—round high table; 12—curved line sofa with high back; 13—curved line track of lights; 14—screen on the wall for cognitive training, physical activity and auto-assessment programmes; 15—composition of two curved line sofas with a table in the middle and a timber screen on the back as a filter for the areas of activities

to make them easily accessible for people coming, and also to guarantee the safety of healthcare professionals, especially during the night.

Create environments for the person's psychological comfort and support. Many of the waiting spaces are designed to promote relaxation and comfort. Multi-sensory environments (MSEs) aim at stimulating all the senses to generate positive sensations and emotions, and to reduce stress, promoting relaxation and recovery [13]. This aim has particular relevance for frail and sensitive users which are common between users of CdC, as described in the previous paragraph, since environment flexibility and customization allow them to self-regulate and rebalance according to their sensory preferences. Flexibility and customization can be provided through design of the environment and its components, but also by automation technologies, allowing a simplified stimulation control. MSE related spaces are typically equipped with sensory devices that provide users with visual, auditory, tactile, olfactory, proprioceptive and even gustatory stimulation.

For this reason, in the project of new waiting areas of CdC Le Piagge two opposite areas are set as multisensory environments (Fig. 4, area A; Fig. 5, sensory room; Fig. 6). These rooms are highlighted by a portal to make the area recognisable and separated from the main corridor to increase privacy, but they are not completely closed to guarantee visibility and permeability of all spaces and consequently safety for patients and caregivers.

These spaces let people customize the space and regulate sensory stimulations to improve comfort and psychological support. The curved shape of the two spaces and the soft furniture (4, 19) gives immediately the idea of welcoming and containment. In these areas every person of every age and different physical and mental conditions can find a comfortable seat and place to stay and wait. As an example, older people can sit on comfortable and homey armchairs (3); children can sit on soft poufs and play with them on the floor (2); parents can lay down on the big curved line sofas and look at their children while they are playing (4, 19); breastfeeding women can sit in the sensory pod (5) in waiting area C and close the curtain to isolate and have more privacy, or go directly in the breastfeeding area in waiting area B (21) that is protected by the wooden filter; in the same way, also a neurodivergent person can use the sensory pod (5) in waiting area C to recreate a sensory balance, have a calm place to stay and relax during waiting time.

Both sensory spaces have natural light and a connection with the exterior and the nature (the green wall—7—and the view of the sky), but can also be darkened by a blind and increase the effect of sensory scenarios. Sensory scenarios include coloured lights, natural and relaxing video projections (8) (on the wall in sensory room in area C and on the ceiling in sensory room in area A), sounds and music.

Through a touchscreen, people are able to choose their favourite scenario to relax, depending on their sensory preferences. As an example, people can choose a scenario that relaxes, with a blue and soft light, the view of a waterfall, the sound of water and

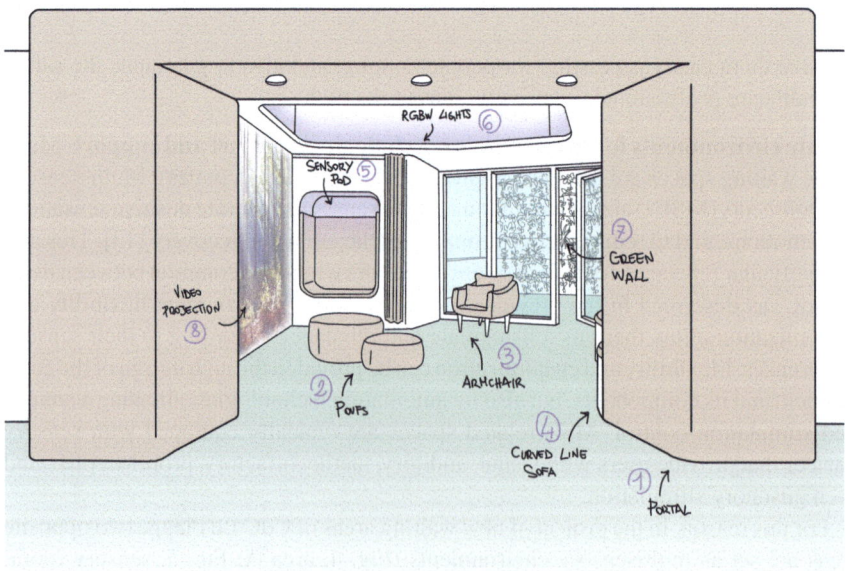

Fig. 6 Representation of the sensory waiting room (area C)

classical music; a dreaming scenario with stars in the sky and a pink colour of light; a scenario that energizes, with running horses, a red light and a strong and louder music; etc. The scenarios can also be composed by the collaboration of healthcare professionals and can be changed during the time, also depending on the different users. Indeed, the sensory areas can be useful to support people with frailties, with neurodiversity or cognitive diseases, or people presenting stress, agitated and anxious behaviours. Changing people's mood and feeling can improve the health experience, but also make them feel more comfortable with healthcare professionals, who often have problems managing people's aggressive behaviours.

Nature is another source of relaxation but also of restoration, as presented in many scientific studies. In the sensory spaces, the contact with nature is provided by projections and sounds, but also by the view of the exterior, especially in the waiting area C, where a green wall (7) is set in front of the main window and the terrace is open towards the interior green courtyard. This window was enlarged in the renovation project to improve the contact with the exterior and to implement the amount of natural light entering. In this way, natural light is also comfortably diffused, as it is indirect and not glaring. This opening gives also the opportunity for people to exit in an open area where they can wait in seats that permit a passive physical activity (9, 10) looking at the green wall or at the garden of the CdC. Moreover, the green wall can be cared for by patients and caregivers to make them directly relate with nature and improve restoration.

Involve, welcome and encourage sociality. The choice of natural materials, such as timber filters (15, 16), warm colours, and homey, comfortable, and curved-line furniture (2, 3, 4, 12, 15, 17), increase the idea of familiarity and welcoming of spaces. Different levels of privacy give the opportunity to choose for relations or individuality, depending on people's preferences.

Anyway, relations are promoted by the orientation of seats and their curved lines to make people look at each other and feel involved in interacting and chatting (Fig. 4). Some of the seats are also movable to give people free control of the space (Fig. 6). A system of sitting (15) is also designed for the corridor areas in all the three waiting spaces, composed of two curved line sofas and a small table in between, to make it possible to sit comfortably and have more privacy, deciding if relating or not with the other people.

Then, the different activities, such as video projections, nature, music, etc., support the relations, to break the ice by sharing the different experiences. Involvement and interaction with spaces are especially encouraged in every area of "action" (see next paragraph) and support the interaction between different people. As an example, in the sensory room in waiting area C (Fig. 6) two older people can sit in the armchairs (3) and start talking looking at children playing on the poufs (2); an older person with the caregiver, or a mom with a children, or a couple, or two people which don't know each other, can sit on the curved line sofa (4) and start making comments about video projections (8), choose together which video or music they want to play, and which colour of light they prefer (6); they can talk about plants outside (7) or about what they are reading on the screen that show health promotion information, etc.

Promote active waiting time. "Active waiting" indicates the possibility of carrying out gentle, short-term physical exercise during breaks, for example in the workplace and/or education in order to interrupt periods of prolonged sedentary behaviour, or during moments of waiting, for example at the bus stop or in a waiting room in order to make the waiting time active [14]. For this reason, in the different waiting areas healthy activities are provided to activate people during waiting time, while sitting, and promote mental and physical health (Fig. 7):

- in the corridor of waiting area C there are two main "stations" of activities. The first "station" is an area dedicated to the development of cognitive training (a). This area is separated and filtered from the corridor and the doors of the medical rooms by a low–high timber filter (15) that hides people when they are seated to increase privacy and make people feel more focused on the training. An armchair makes it possible to comfortably sit in front of a screen (14) where different programmes/games train people's minds. This system makes it possible to prevent cognitive diseases and to inform people to enhance knowledge about mental diseases. Programmes and games will be developed in collaboration with the healthcare professionals to be addressed to their patients and caregivers;

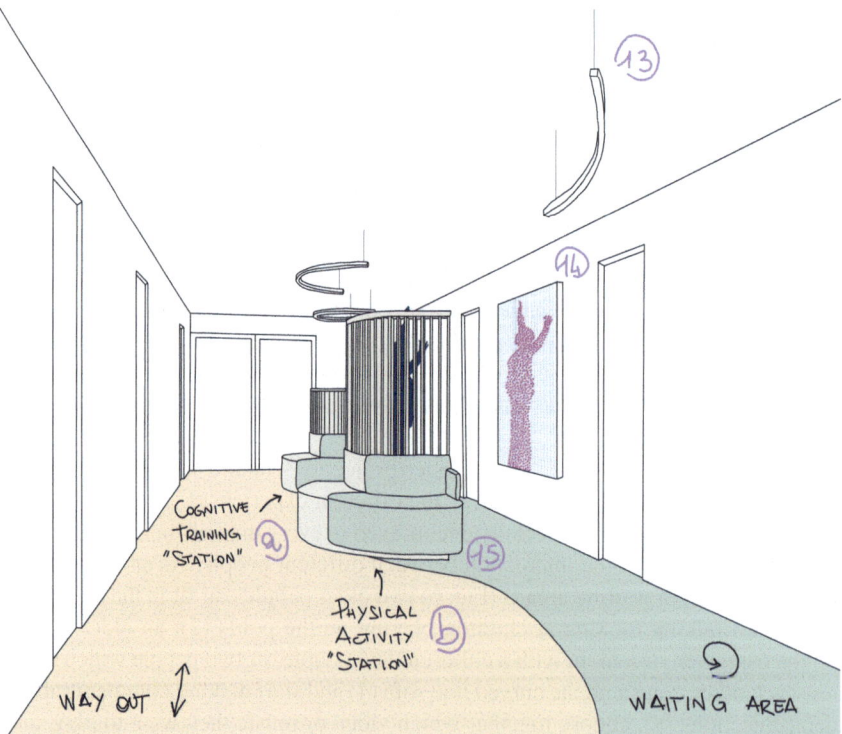

Fig. 7 Representation of the active waiting corridor (area C)

- in the same corridor of waiting area C, the second "station" is dedicated to the promotion of movement and gentle short-term physical activity (b). The area is separated and filtered from the corridor and the doors of the medical rooms by a high timber filter (15) that hides people when they are standing. People will be in front of a screen (14) where different programmes/games make them move and promote physical activity. This makes it possible to prevent chronic diseases and to enhance knowledge about them. The prevention is possible also by an auto-assessment system that makes evidence of a possible disease by some exercises, as detailed in Bertiato et al. [15]. These results can be used by patients and caregivers to improve knowledge but also to be shared with healthcare professionals and improve digital health;
- two areas for children to play (18), visible by the sensory area in waiting area A (Fig. 4) and from the breastfeeding area in waiting area B (Fig. 4) where parents can stay and relax as their children are in a safe and dedicated environment;
- the terrace in waiting area C (Fig. 5), an outdoor and covered area where it is possible to wait independently of weather conditions. In this area it is possible to sit on special furniture (9, 10) which makes people do passive training, as explained in Bertiato et al. [15], improving people's physical health and active ageing. In this area, it is also possible to take care of plants on the green wall (7), as explained in the paragraph about comfort and restoration, making people exit, stand and move during waiting time. The plants can also represent a positive memory for people, especially for older people and people with chronic diseases who are going to be frequent users of the CdC and improve positive feelings in coming back.

Promote and activate health. All the different waiting areas use technological systems and devices to improve information and increase knowledge on health and on CdC services. In this sense, the technologies aim at supporting health promotion by facilitating the sharing of information. As an example, while people are sitting health messages can pass on screens and make them curious about healthy life-styles and behaviours and also encourage them to use the health "stations" spread in the different waiting areas. As explained in the previous paragraphs, physical activity is promoted both as passive training or active movement, and facilitated by the configuration of the space, special furniture and integrated technologies. In the same way, mental health is promoted by cognitive training "station" by programmes and games which involve people during waiting time, but also by the configuration of comfortable and welcoming spaces and the interaction with the two sensory environments that improve psychological comfort and support. In the programmes/games of health "stations" (both physical activity and cognitive training areas) self-assessments are also provided to favour prevention of chronic pathologies and knowledge of physical and mental diseases.

4 Conclusion

This project makes evidence of some key points in the design of healthcare environments:

- the possibility to do a few actions to transform the existing healthcare facilities and promote people's health and wellbeing;
- the possibility to improve a typical healthcare rigid and anonymous layout (e.g., a long straight central corridor overlooked by the outpatient rooms), even if it is already configured;
- the development of innovation in waiting spaces: transforming waiting into active waiting time while doing different activities which promote healthy and active ageing;
- the affirmation of the value of designing according to Design Guidelines which are based on a rigorous scientific research process.

Competing Interests This study was developed within the project funded by Next Generation EU—"*Age-It*—Ageing well in an ageing society" project (PE0000015), National Recovery and Resilience Plan (NRRP)—PE8—Mission 4, C2, Intervention 1.3.

The views and opinions expressed are only those of the authors and do not necessarily reflect those of the European Union or the European Commission. Neither the European Union nor the European Commission can be held responsible for them. This contribution presents some results of the research of Task 1.2—Design strategies to improve active/healthy ageing in primary healthcare facilities, led in Spoke 9—Advanced Gerontechnologies for active and healthy ageing.

The new waiting spaces of Casa della Comunità Le Piagge were developed by the collaboration of a multidisciplinary team which involved the researchers of TESIS center, Department of Architecture, University of Florence; the company GLM Engineering, the healthcare professionals of CdC Le Piagge and the Health Trust of Florence (AUSL Toscana Centro).

The authors have no conflicts of interest to declare that are relevant to the content of this chapter.

References

1. Evans, G. W., & McCoy, J. M. (1998). When buildings don't work: The role of architecture in human health. *Journal of Environmental Psychology, 18*(1), 85–94.
2. Del Nord, R. (Ed.). (2006). *Lo stress ambientale nel progetto dell'ospedale pediatrico: Indirizzi tecnici e suggestioni architettoniche*. Motta Architettura.
3. von den Bosch, M. (Ed.). (2018). *Oxford textbook of nature and public health: The role of nature in improving the health of a population*. Oxford University Press.
4. Kaplan, R., & Kaplan, S. (1989). *The experience of nature: A psychological perspective*. Cambridge University Press.
5. Ulrich, R. (1984). View through a window may influence recovery from surgery. *Science, 224*(4647), 420–421.
6. Ulrich, R. S., Simons, R. F., Losito, B. D., Fiorito, E., Miles, M. A., & Zelson, M. (1991). Stress recovery during exposure to natural and urban environments. *Journal of Environmental Psychology, 11*(3), 201–230.
7. Verderber, S. (2010). *Innovations in hospital architecture*. Routledge.

8. International Network of Health Promoting Hospitals and Health Services. (2020). *Global HPH strategy 2021–2025*. International HPH Network. https://www.hphnet.org/wp-content/uploads/2021/02/Global_HPH_Strategy-2021-2025.pdf
9. Golembiewski, R. T. (2017). Salutogenic architecture in healthcare settings. In M. Mittelmark (Ed.), *The handbook of salutogenesis* (pp. 267–276). Springer International Publishing.
10. Allen, J., Bernstein, A., Cao, X., Eitland, E. S., Flanigan, S., Gokhale, M., Goodman, J. M., Klager, S., Klingensmith, L., Laurent, J. G. C., & Lockley, S. W. (2017). *Building evidence for health: The 9 foundations of a healthy building*. Chan School of Public Health.
11. Del Nord, R., & Peretti, G. (2011). *L'umanizzazione degli spazi di cura. Linee guida*. Ministero della Salute, Tesis.
12. Ferrante, T., Setola, N., Villani, T., Bellini, E., & Romagnoli, F. (2024). Innovazione e cura: La progettazione di spazi smart per l'assistenza agli anziani. *Progettare per la Sanità, 1*, 32–41.
13. Bellini, E., Macchi, A., Setola, N., & Lindahl, G. (2023). Sensory design in the birth environment: Learning from existing case studies. *Buildings, 13*(3), 604.
14. Muñoz-Parreño, J. A., Belando-Pedreño, N., Torres-Luque, G., & Valero-Valenzuela, A. (2020). Improvements in physical activity levels after the implementation of an active-break-model-based program in a primary school. *Sustainability, 12*(9), 2–12.
15. Bertiato, F., Bellini, E., & Setola, N. (2024). KALI project: A new concept for health promotion within waiting rooms. In L. Fiorini, A. Sorrentino, P. Siciliano, & F. Cavallo (Eds.), *Ambient assisted living. ForItAAL 2024. Lecture notes in bioengineering* (pp. 30–50). Springer.

Open Access This chapter is licensed under the terms of the Creative Commons Attribution 4.0 International License (http://creativecommons.org/licenses/by/4.0/), which permits use, sharing, adaptation, distribution and reproduction in any medium or format, as long as you give appropriate credit to the original author(s) and the source, provide a link to the Creative Commons license and indicate if changes were made.

The images or other third party material in this chapter are included in the chapter's Creative Commons license, unless indicated otherwise in a credit line to the material. If material is not included in the chapter's Creative Commons license and your intended use is not permitted by statutory regulation or exceeds the permitted use, you will need to obtain permission directly from the copyright holder.

Adaptability of Housing for Home Care: Strategies and Design Solutions for Assistive Spaces and Technologies

Federica Romagnoli, Teresa Villani, and Tiziana Ferrante

Abstract The growing demand for home care among older adults calls for a rethinking of design strategies to make the home suitable as the "primary place of care." This contribution presents the initial results of a research project conducted within the Age-It program, focused on the adaptability of domestic environments to home care activities and the potential integration of assistive technologies. Based on a person-centered and bio-psycho-social model, the study analyzes the interaction between frail users, caregivers, and the living space, with the aim of identifying the spatial and technological requirements necessary to support care activities in an integrated manner. The research methodology began with an analysis of user needs, with the goal of identifying environmental and technological requirements that characterize spaces used for home care. The outcomes of this process led to the definition of operational guidelines to support the design of adaptations in the areas most involved in care activities. These include recommendations for ensuring safety, comfort, usability, maintainability, and the integrability of assistive technologies, as well as the proposal of alternative design solutions for adaptation, which are qualitatively assessed based on their construction impact and compatibility with existing constraints. Such solutions aim to address the needs of both care recipients and caregivers in an integrated way, while promoting ageing-in-place even in conditions of reduced self-sufficiency. The study highlights the importance of a proactive and multidisciplinary approach to designing "care-ready" environments capable of responding to the evolving housing and health needs of the population.

Keywords Home adaptation · Design for home care · Design for health · Ageing in place · Ambient assisted living

---

F. Romagnoli (✉) · T. Villani · T. Ferrante
Department of Planning, Design, Architectural Technology, Sapienza University of Rome, Rome, Italy
e-mail: federica.romagnoli@uniroma1.it

© The Author(s) 2025
T. Ferrante and M. Sacco (eds.), *Habitable Future*,
SpringerBriefs in Applied Sciences and Technology,
https://doi.org/10.1007/978-3-031-95735-2_12

1 Introduction

1.1 The Adaptability of the Home as the 'Primary Place of Care'

The increase in life expectancy and the growing incidence of chronic-degenerative diseases among older adults call for a comprehensive rethinking of local health and social care services, as well as the physical spaces in which such services are delivered. In particular, the recent reforms introduced through Italy's National Recovery and Resilience Plan (PNRR), enacted by Law No. 33/2023, designate the home as the "primary place of care", promoting a model of de-institutionalization based on proximity, continuity of care, and the personalization of services—especially for older individuals who are no longer self-sufficient and require ongoing, long-term assistance.

Within this framework, the research carried out as part of the Age-It program—Task 1.3 of Spoke 9—aims to investigate the relationship between non-self-sufficient elderly individuals, caregivers, and the domestic environment, with the goal of developing strategies for adapting housing to home care activities. To this end, an intermediate objective of the study was the definition of design guidelines, based on scientific evidence and best practices, capable of guiding interventions for adapting the physical environment in an age-friendly way, integrating Ambient Assisted Living (AAL) technologies.

The research is aligned with a human-centered [1] and bio-psycho-social [2] model, which considers the interaction between the individual and the environment as a fundamental element in ensuring well-being and the quality of care (Fig. 1).

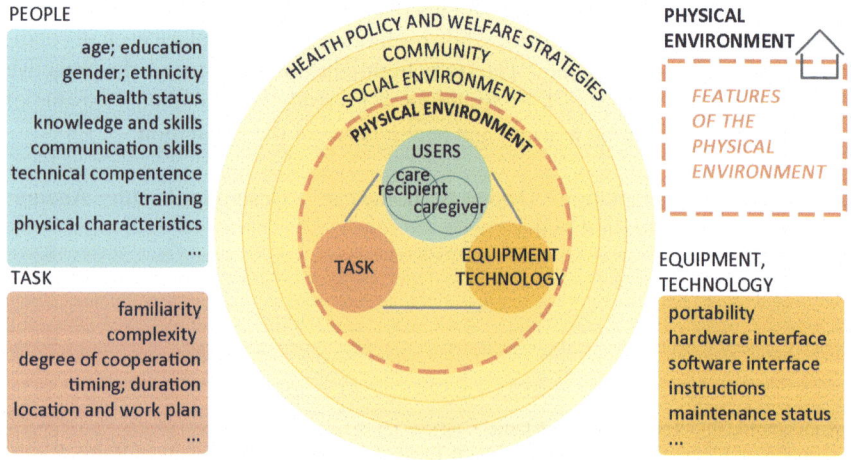

Fig. 1 The human-factors model of home care. Adapted by the authors from [1]

1.2 Research Phases and Activities

Building upon the human-centered model, the research adopted a demand-performance-based approach. The main user groups involved—namely, older adults receiving care and their caregivers—were identified, and their needs were interpreted and subsequently translated into spatial features (*environmental requirements*) and features of physical elements (*technological requirements*), such as walls, furniture, and devices that configure the domestic space. These features are considered essential for enabling the home environment to support daily living and care activities in an integrated manner.

In line with this approach, the research was structured into the following phases.

In Phase 1, "Definition of the needs framework for older adults receiving home care and their caregivers," qualitative and quantitative information was collected on the functioning of home care services in Italy. This was achieved through the analysis of statistical and regulatory sources, interviews with healthcare professionals and caregivers, direct observations, and the administration of structured questionnaires. The process of systematizing and interpreting this information led to the definition of the users' needs framework—both for care recipients and caregivers—in relation to the care activities carried out within the home environment.

In Phase 2, "Definition of housing requirements for home care," the state of the art was reconstructed regarding the design principles of Design for Ageing in Place and Universal Design, as well as assistive Ambient Assisted Living (AAL) technologies that support care provision.

In parallel, case studies of Health Smart Homes were analyzed in order to identify the main strategies for integrating assistive devices within the domestic environment. These activities led to the development of a structured framework of environmental and technological requirements for the spaces (*environmental units*[1]) most directly involved in home care activities. The identified requirements subsequently informed the structuring of the design guidelines, which define the criteria for guiding decision-making in housing adaptation projects.

Finally, in Phase 3—currently under development—the proposed guidelines will undergo validation. Their application to pilot cases will be supported by a design tool for housing adaptation based on Building Information Modeling (BIM) methodologies. This tool, developed in accordance with the proposed guidelines, is intended to assist designers in simulating and evaluating alternative intervention strategies within virtual prototypes of specific domestic environments (Fig. 2).

The outcome of these activities will be an integrated system of knowledge, design strategies, and innovative project management tools aimed at making homes more inclusive, safe, and adaptable to the needs of older adults and their caregivers, in line with a multidisciplinary vision of the challenges posed by home care.

---

[1] The Italian technical standard UNI 10838:1999 defines an "environmental unit" as "a grouping of activities (…) that are spatially and temporally compatible with one another" (*translation adapted by the authors*). In this paper, the term refers to the various rooms or spaces that comprise a home.

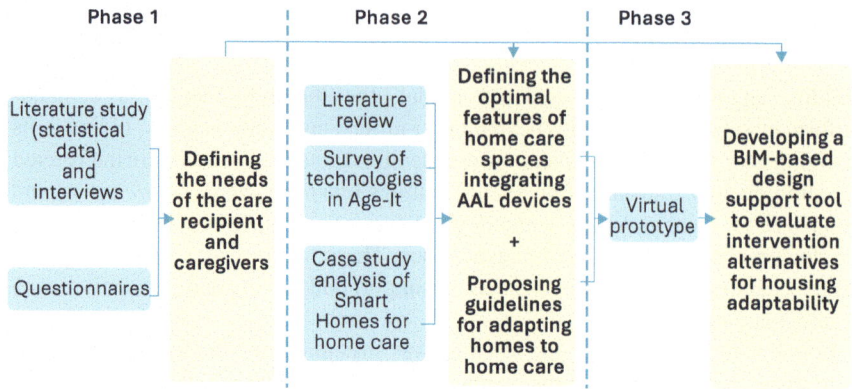

Fig. 2 Research phases and activities

2 Home Care in Italy: Users, Activities and Needs

2.1 Organization of Home Care Services in Italy

In Italy, Integrated Home Care (ADI) represents one of the main responses to the growing demand for healthcare among older adults, particularly in relation to chronic-degenerative diseases and conditions of non-self-sufficiency. ADI is structured as a system of coordinated medical, nursing, and rehabilitative interventions, integrated with social services, and aimed at supporting vulnerable individuals within their own homes. This model places the person at the center of a territorial network of healthcare services. The primary objective is to reduce hospital admissions by promoting the delivery of treatments, assistive devices, medications, and consultations directly at the patient's home, also through the use of telemedicine and remote monitoring technologies.[2]

2.2 Users and Their Spatial Needs

Older adults receiving care. In Italy, non-self-sufficient older adults account for 82% of home care service recipients [3]. Their profile is characterized by advanced age (average of 83 years), a higher prevalence of women (66%), and a significant incidence of social isolation (58% live alone) and dependence in daily activities (41%) [4]. The most common chronic conditions include heart failure, ischemic heart disease, dementia, cerebrovascular diseases, and respiratory illnesses [5]. Among these, heart failure is particularly significant due to its high clinical complexity,

---

[2] cf. Italian Ministry of Health, Decree of 21 September 2022—"Guidelines for telemedicine services—Functional requirements and service levels".

requiring continuous treatment and frequent use of medical and assistive technologies. These conditions increase the likelihood that home adaptation interventions will be necessary [6]. The progression of heart failure often leads to cognitive impairment, multiple comorbidities, and geriatric syndromes such as immobility and pressure ulcers [7], which require intensive and frequent home-based care.

In this context, the home becomes not only a place of everyday life, but also a therapeutic environment. The spatial characteristics of the dwelling must respond to the psycho-physical needs of the care recipient who, even in cases of severe functional limitations, may retain some forms of autonomous activity that should be encouraged through stimulating and comfortable environments.

These activities may include, for example: sensory interaction with the surrounding environment (looking, listening, touching); rest and privacy (relaxing, sleeping); and light occupational activities such as listening to music, reading, or manipulating objects.

All of these require environments that are well-lit, acoustically insulated, and thermally comfortable, designed with pleasant-to-touch materials and furnishings that are both care-supportive and domestically styled—avoiding the transformation of the home into a clinical or overly medicalized setting.

I caregiver. Within the human-centered model, caregivers are fully recognized as users of the home care system.

Although the needs of the different categories of caregivers are highly diverse—including healthcare professionals such as physicians, nurses, and physiotherapists; social care providers such as personal care assistants and home aides; and informal caregivers such as relatives or volunteers—they all share the burden of activities that demand considerable physical effort and psychological engagement. This makes the design of ergonomic, accessible, and well-equipped spaces essential. Such spatial considerations must take into account the most frequent activities carried out in the home setting, including: mobilization and hygiene care of the care recipient, administration of therapies, clinical monitoring (including remote monitoring), maintenance of environmental hygiene, psychological support, and communication with both the person receiving care and other caregivers [8].

To better identify the most critical issues in the relationship between care activities and domestic space, the research group developed and administered a structured questionnaire to 76 professional caregivers (including physicians, nurses, and physiotherapists) and 26 family caregivers. The results showed that 77% of care activities take place in the bedroom; 45% of caregivers do not have access to a convenient, functional surface near the person receiving care; and in 32% of cases, care equipment remains in the home, mainly in the room dedicated to care activities. The main issues reported by participants included: the presence of architectural barriers (such as level changes and narrow passageways); insufficient bedroom space, which limits the use of adjustable beds and mobility aids; inaccessible bathrooms, often inadequate for assisted personal hygiene; and a lack (or insufficiency) of storage space and cabinets for medical equipment and consumables.

The user needs framework. The set of exploratory activities described above led to the definition of the user needs framework, summarized in Table 1. This framework represents the collection of core information that was interpreted and systematized in order to define the optimal housing requirements for home care.

It expresses the "demands" to which the physical space must respond in order to be suitable for hosting care activities in an integrated manner alongside everyday living.

3 Design for Adaptability and Home Care: Principles, Barriers, and Opportunities

The increasing demand for home care services for non-self-sufficient older adults makes it urgent to adapt the existing housing stock, which remains largely inadequate in meeting the stringent requirements of accessibility, safety, and comfort associated with home care settings. According to recent studies, approximately 80% of homes in Europe do not meet the needs of older populations [9], and only a small portion is effectively suited to accommodate medical and assistive devices [10].

The concept of *home modification* or *home adaptation*, introduced in the 1990s, refers to targeted interventions aimed at transforming the domestic environment to improve individual autonomy and safety, while also reducing caregiver burden and the risk of injury [11].

In Italy, Ministerial Decree 236/1989 defines adaptability as the "possibility of modifying built spaces over time, at limited cost,[3] to render them fully usable even by persons with reduced motor or sensory capacities."

At the international level, the principles of Design for Adaptability outlined in ISO 20887:2020 recommend adherence to Universal Design principles in order to accommodate future, yet unknown, user needs, thereby avoiding costly transformations.

A *life-span* design approach—which considers the evolving needs of individuals over time—is now essential not only for new construction but also, and above all, for increasing the adaptability of the existing housing stock, especially considering that 80% of homes projected to meet demand through 2050 have already been built [12].

However, the process of adapting existing buildings encounters numerous political, cultural, economic, and technical barriers [13]. Additionally, the stigmatizing perception associated with design solutions that resemble clinical or hospital-like settings often impedes the acceptance of domestic transformations [14].

It is therefore essential to plan in advance for spaces to be care-ready, in order to prevent critical issues and enhance the integration of furnishings and equipment

---

[3] That is, through the deferred execution of works over time that do not alter the load-bearing structure or the common utility systems.

Table 1 Users, activities, needs, and relationship with physical space as a reference framework for home adaptation

	Users	Activities	Needs	Relationship with physical space
CARE RECIPIENT	Non-self-sufficient older adult with chronic conditions (heart failure)	• Sensory interaction with the environment • Rest and maintaining personal privacy • Occupational and recreational therapy	• Enjoy tranquility, rest, and privacy (acoustic comfort) • Ability to control and interact with the environment through multiple senses • Ability to regulate thermal-hygrometric conditions (Comfort/Management) • Ability to adjust lighting levels (Visual Comfort) • Benefit from positive distractions (e.g., views of green spaces) • Be surrounded by furnishings and devices with a domestic appearance • Preserve the natural circadian rhythm of wake and sleep (Visual Comfort) • Perform rehabilitative and occupational exercises (Usability)	• Properly lit, quiet, and thermally comfortable environment (**Bedroom**) • Quiet and private room with full blackout capability (**Bedroom**) • Presence of specific equipment and environmental features for lighting, noise control, and temperature (**Bedroom; Bathroom**)

(continued)

Table 1 (continued)

	Users	Activities	Needs	Relationship with physical space
CAREGIVER	Healthcare and social care professionals, family members, home care assistants, volunteers, etc.	Assistance with: • Mobilization/deambulation/rehabilitative treatments • Personal hygiene • Feeding/nutritional treatments • Therapy monitoring and space management • Diagnostic evaluation • Wound care and treatment of tissue alterations • Respiratory function treatments	• Perform care activities safely (Safety) • Have maneuvering space around the care recipient's bed (Usability) • Benefit from mobility aids for assisting the care recipient (Usability/Management) • Use assistive technologies for health monitoring (Management/Integrability) • Regulate thermal and hygrometric conditions (Comfort/Management) • Use appropriate lighting for precision tasks and diagnostics; adjust lighting levels as needed (Visual Comfort) • Have support surfaces near the care recipient for placing equipment (Usability/Management) • Store equipment after treatments (Usability/Management) • Maintain cleanliness of spaces with ease (Management) • Ensure suitable lighting and acoustic conditions for telemedicine consultations (Visual/Acoustic Comfort)	• Wide maneuvering spaces passageways and clearances free of obstacles considering appropriate equipment (**Entrance; Bedroom; Bathroom**) • Proper microclimate, acoustic, and lighting (**Bedroom; Bathroom**) • Clean and hygienic environment; removable support surfaces; (**Entrance; Bedroom; Bathroom**) • Spatial configuration that supports monitoring (**Bedroom; Bathroom**) • Availability of cabinets/storage for devices and equipment (**Entrance; Bedroom; Bathroom**)

within the home, while promoting universal and inclusive solutions that preserve the residential character of the environment.

Most of the methods and tools currently available for designing or assessing the adaptability of buildings are based on the theories of S. Brand, who argued that a building's adaptability is inversely proportional to the level of interdependence among its parts—namely, site, structure, envelope, systems, spatial configuration (such as partitions, fixtures, and flooring), and furnishings—which are subject to replacement or modification at different intervals [15].

The adaptability of existing housing remains a complex issue, for which no standardized procedures are currently established [16]. Structural, technical, and regulatory constraints represent significant barriers to feasible interventions, as do constraints related to the historical and artistic heritage, which is particularly relevant in the Italian context.

In Italy, domestic adaptation is regulated by Law No. 13/1989, which provides financial incentives and tax deductions for building works requested by individuals with disabilities or reduced self-sufficiency. In some Italian regions, Domestic Environment Adaptation Centers (CAAD) have been established to act as intermediaries between institutions and citizens, offering consultancy and design services for accessible and adaptable housing.

In synergy with existing measures, it is clear that adopting a preventive approach aimed at preparing homes to be "care-ready" would make it possible to anticipate the spatial challenges that typically emerge only when specific care needs arise. This would help to avoid invasive construction works or the temporary relocation of the care recipient to a healthcare facility.

Furthermore, if supportive design solutions capable of accommodating age-related or health-related fragilities are integrated during renovation processes, they can significantly enhance the long-term usability and value of the home, considering the evolving needs of its occupants over time.

4 A Proposal for Guidelines on Designing Adaptable Spaces for Home Care

4.1 Objectives and Structure of the Guidelines

The complexity of the challenge of adapting housing to home care highlights the need to identify, systematize, and implement design support tools that can facilitate the application of life-span design principles and the integration of AAL technologies into domestic environments. This is made possible by identifying the most appropriate characteristics of technical elements, finishes, and furnishings.

To this end, the guidelines proposed within this research aim to guide design choices and support shared decision-making between professionals and users—namely the non-self-sufficient older adult and their caregiver—in the adaptation of the domestic environment.

The development of the guidelines was based on exploratory and analytical activities (including the analysis of best practices, literature reviews, and studies on the integrability of AAL devices) specifically focused on home care activities related to older adults with heart failure.

Starting from the previously defined user needs framework (see Sect. 1.2), the analysis of the needs of care recipients and caregivers was deepened in relation to the specific environments (*environmental units*) most relevant to home care practices.

For each of these environmental units, key characteristics were identified—such as safety, usability, indoor environmental comfort, ergonomic design, and furnishability—along with corresponding reference parameters (*environmental requirements*).

These were then used to determine the expected performance levels (*technological requirements*) related to technical elements (e.g., walls, windows, etc.) as well as assistive devices and furnishings.

This set of requirements forms the foundation of the proposed guidelines and serves as a reference framework to guide technical design decisions within housing adaptation projects.

Given the high variability in the contextual conditions of existing homes, it was necessary to formulate an operational abstraction of potential interventions, based on recurring situations. The proposed interventions described in the guidelines were selected not only through in-depth case study analysis and literature review, but also through market research on emerging AAL systems, prioritizing low-impact technical solutions that meet criteria for safety, accessibility, and environmental sustainability.

The guidelines were structured according to the following criteria:

- to satisfy all the requirements—namely, safety, comfort, usability, management, and integrability—considered essential for the suitability of spaces to host home care activities;
- to take into account technical (structural and system-related) and/or regulatory constraints;
- to minimize the impact of the construction work involved;
- to limit costs and resource consumption;
- to promote the use of environmentally low-impact, flexible, adaptable, and reversible technical solutions.

Starting from a set of general recommendations that consider the home as a whole, the guidelines then focus on the spaces (*environmental units*) identified as most relevant to home care3F[4]: the entrance, the care recipient's bedroom, and the bathroom (Fig. 3).

---

[4] See the results of the survey described in Sect. 2.2.

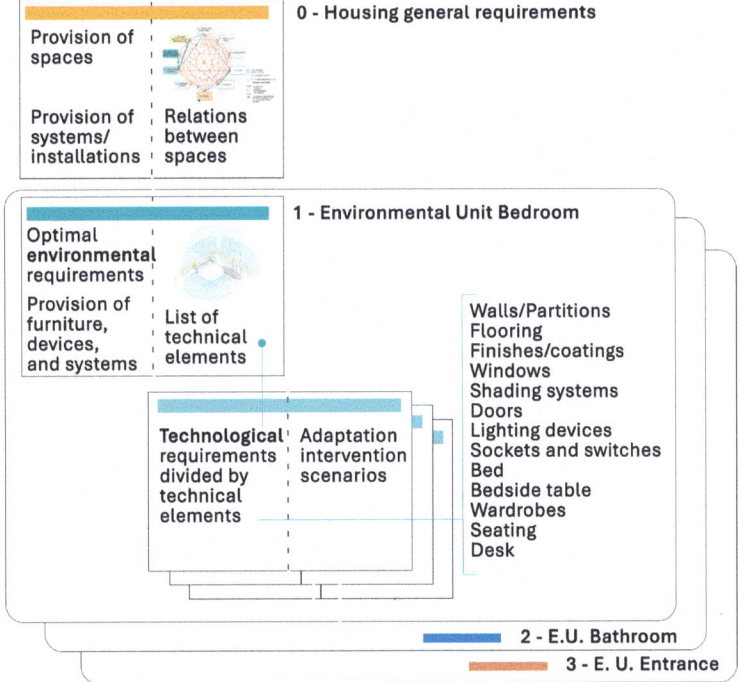

Fig. 3 Structure of the guidelines

The first section of the guidelines provides recommendations on the minimum desirable dimensions of spaces, the ideal configuration of infrastructure systems, and the optimal spatial relationships and proximity between rooms involved in home care activities.

The aim is to support an initial optimal definition of possible home reconfigurations.

In particular, the guidelines offer suggestions regarding:

- direct spatial connections between the entrance, the care recipient's bedroom, and the bathroom, in order to avoid interference with the privacy of other household members;
- the width and accessibility of interior pathways, free from barriers such as thresholds or steps;
- the availability of storage space for assistive devices and medical equipment, which should be easily accessible but discreetly placed to preserve the domestic character of the home;
- adjacency between the care recipient's bedroom and that of the caregiver, to facilitate monitoring and enable a rapid response in case of emergency.

The remaining three sections of the guidelines focus on the three environmental units identified as most relevant to home care: the entrance, the care recipient's bedroom, and the bathroom.

For each of these spaces, key requirements have been identified and classified according to the six categories defined by the Italian technical standard UNI 8289: *Safety, Comfort, Usability, Appearance, Management*, and *Integrability* (see Sect. 4.2).

Specifically, as will be described in the following sections, for each environment the following contents have been developed:

- **a summary of environmental requirements and the optimal provision of furniture, devices, and technical systems, compiled within a dedicated sheet that includes (see Sect. 4.3):**
  - a structured list of environmental requirements, referring to dimensional characteristics and indoor comfort parameters, classified according to the six categories defined by UNI 8289;
  - the ideal provision of furniture and assistive devices;
  - the ideal provision of technical systems;
  - a schematic representation of the environment, highlighting the technical elements, furnishings, and devices that will be further analyzed in dedicated in-depth sheets;

- **a set of in-depth technical sheets focusing on individual building elements and their corresponding technological requirements, presented in a structured format that includes (see Sect. 4.4):**
  - a list of technological requirements, referring to the expected performance of the examined element, classified according to the six UNI 8289 categories;
  - a set of proposed adaptation alternatives, qualitatively assessed based on the impact of the involved construction work.

4.2 Key Characteristics of Spaces for Home Care

The following section describes the characteristics of domestic environments by relating recurring user activities, specific needs associated with home care, and the corresponding spatial requirements (classified according to the UNI 8289 standard). These relationships are used as a foundation for developing design recommendations aimed at enhancing the adaptability of the home.

Safety. In all spaces used for home care, particularly considering potential mobility impairments related to health conditions, it is essential to ensure the absence of obstacles that may lead to falls. Glossy or polished flooring should be avoided to reduce the risk of slipping and to facilitate the identification of any liquids on the floor—especially in areas where water or therapeutic solutions are frequently used. Surfaces with complex patterns or high-contrast designs should also be avoided, as they may conceal small objects and increase the risk of tripping. In high-risk areas, such as the shower or sink area, the use of non-slip flooring is recommended. An additional safety measure could be the use of shock-absorbing flooring capable of reducing injuries in the event of a fall. Other important precautions include outward-opening doors and thermostatic valves to control the temperature of domestic hot water.

All areas should be well lit, with switches placed in strategic locations—at the beginning and end of frequently used pathways—and clearly visible even in the dark. To prevent impact-related injuries, furnishings should always be free of sharp edges. For fire safety, it is essential that surface finishes and furnishings have a low reaction to fire. The presence of smoke detectors, connected sensors, and automated alarm systems represents a crucial technological contribution to improving overall protection.

Visual comfort. The quality of lighting is closely linked not only to safety during care-related activities but also to the psychological well-being of the care recipient. In all rooms, the daylight-to-floor area ratio should exceed the minimum regulatory thresholds, and artificial lighting should be adjustable—both diffuse and task lighting (preferably using mobile and/or directional fixtures). Special attention must be paid to the positioning of light sources to avoid interference with the visual field of bedridden users. It is equally important to be able to shield direct sunlight, especially during the summer months, using adjustable shading and blackout systems. These systems should ensure both privacy and the ability to achieve total darkness when needed, which is essential for rest. Automated lighting control systems, connected to timers or sensors, can further enhance visual comfort while optimizing energy consumption.

Particular care should be taken in the selection of finishes and coverings with high reflectance values. Neutral and light-colored tones are preferred to ensure good distribution of natural light and to facilitate the recognition of the skin tone of the person receiving care. Conversely, mirrored surfaces should be avoided as they can cause glare, along with complex decorative patterns that may interfere with visual perception.

The quality of views also plays a significant role in promoting well-being. Views of vegetated open spaces or the presence of indoor plants help reduce stress and support the regulation of circadian rhythms [17]. It is important to position the bed so that the person, whether seated or lying down, has a direct line of sight to the outside, which can help prevent depressive states and disorientation.

At the same time, privacy must be ensured by preventing views from other rooms, and medical equipment should be positioned outside the user's visual field in order to maintain a non-institutional, home-like atmosphere.

From a perceptual standpoint, furnishings, finishes, and colors should convey a sense of warmth and well-being, and should not resemble a clinical setting. For example, support elements may be integrated or concealed within traditional furniture, applying principles of biophilic design and its regenerative effects through the use of natural materials [18].

Thermal and hygrometric comfort. As physiological thermoregulation capacity decreases with age, it is essential to ensure an ambient temperature that is both appropriate and easily adjustable. Preventing the formation of unwanted air currents is equally important, which can be achieved through the careful positioning of openings and ventilation outlets, as well as by using automated or remotely controlled systems for managing windows and ventilation, connected to temperature and air quality sensors.

Acoustic comfort. Silence and insulation from external noise are key factors in promoting the well-being of the care recipient. For this reason, the bedroom should be free from sources of noise, thus supporting a calm environment—also through remote monitoring technologies, which reduce the need for active supervision by caregivers, as this could itself generate further noise.

The use of sound-absorbing materials in furnishings, wall coverings, or suspended ceilings also contributes to improving the acoustic quality of the environment, facilitating the comprehension of spoken language.

Olfactory well-being and indoor air quality. To ensure proper indoor air quality—crucial for the health and well-being of the care recipient—it is necessary to limit exposure to outdoor pollutants and harmful substances emitted by paints, finishes, and furniture, while also preventing the accumulation of humidity and mold formation. For this purpose, materials free from formaldehyde and volatile organic compounds (VOCs) should be selected, and effective air exchange should be promoted through natural ventilation. This can be supported by double-facing rooms and openings proportionate to the room's surface area. Window systems should allow for multiple opening modes in order to regulate airflow according to environmental conditions and user needs. Technological solutions such as air purification systems can also contribute to maintaining a healthy indoor environment.

Tactile comfort. To foster feelings of comfort, well-being, and perceived safety for the care recipient, the selection of surface materials (including textiles and furniture finishes) should take into account their tactile qualities. This may include evaluating characteristics such as hardness, texture (smoothness or roughness), and thermal conductivity, which influence how the material feels to the touch and the temperature it conveys on contact.

Usability of spaces and furnishings. In the context of home care, it is essential to go well beyond standard accessibility regulations, paying particular attention, for example, to the ease with which users can open doors and windows. This includes considering the ergonomic shape of handles and the force required to operate them, as well as the use of ergonomically designed switches that are easy to activate.

At the entrance, sufficient space should be provided to allow healthcare workers to sit and change their clothes comfortably.

In the bedroom, it is essential to ensure access to both sides of the bed, to reserve a sterile area for medical equipment, and to allocate adequate space for machines and containers for special medical waste. The room should also include movable support surfaces and sufficient shelving to store assistive devices and consumable materials, while avoiding clutter at floor level. A flexible furniture layout should also be considered in the bathroom—for example, by using removable storage units beneath the sink to allow wheelchair users to approach it easily.

Finally, it is important to provide multisensory environmental cues (e.g., color contrasts) to ensure that spaces and furnishings are easily recognizable, even for individuals with reduced visual or cognitive abilities.

Management—cleaning and maintainability of spaces. All surfaces involved in home care activities must be easy to clean, wear-resistant, and non-porous, in order to reduce the risk of contamination. Durable materials compatible with disinfectant cleaning agents should be used, while materials prone to degradation or that trap allergens—such as carpeting or rugs—should be avoided.

Continuous flooring (without visible joints), antistatic, and abrasion-resistant surfaces are particularly suitable in areas subject to heavy use or the movement of wheeled equipment. Vertical surfaces should also be washable and impact-resistant, with corners and joints designed to minimize the accumulation of dust.

Integrability and adaptability of technical elements. The integration of medical devices and assistive technologies into the domestic environment requires particular attention to the functional and infrastructural flexibility of building systems. These systems must allow for extensions, modifications, inspections, and maintenance without the need for invasive demolition. To ensure the proper functioning of fixed equipment, the home should be equipped with a suitably dimensioned electrical system, including a sufficient number of dedicated outlets and service continuity provided by backup batteries or generators.

It is also essential to ensure the presence of Internet connections—wired or wireless—to support the installation and operation of environmental sensors via stable and secure data networks.

To this end, the use of technical counter-walls or suspended ceilings, as well as accessible conduits in strategic locations, is recommended. These solutions facilitate the placement of medical equipment and the installation of home automation systems (such as for doors, windows, shading devices, etc.).

To increase the spatial flexibility of domestic environments (e.g., to add a room for a caregiver), it is advisable to facilitate the transformability of interior partitions using dry-assembled systems [19], and to pre-engineer service connections to support the merging or subdivision of spaces [20, 21].

An effective system infrastructure should also include a dedicated, inspectable technical compartment for housing servers used in the management of AAL systems.

Finally, walls and floors should be equipped or structurally prepared to support fixed installations, including assistive technologies and systems for lifting and transferring user.

Device Characteristics and Integrability. The push toward technological innovation is also introducing a new paradigm in the reconfiguration of domestic spaces intended for home care.

As part of the guidelines, for each of the rooms under examination, optimal sets of assistive devices and technologies have been identified and classified according to ISO 9999:2022. Where relevant, standard dimensions are also provided, in order to support the selection of integration strategies adapted to contextual conditions. The list of devices—organized by requirement categories and key characteristics relevant to their integrability—can serve as a guide for designers in selecting solutions compatible with the domestic spaces to be adapted.

For each device or assistive technology, special attention should be paid to the placement of any necessary sensors, the aesthetic integration of the devices (whether visible or embedded within objects or technical elements), and the environmental conditions required for their operation (e.g., temperature, humidity, lighting, noise levels, etc.).

4.3 Environmental Requirements of Spaces and the Provision of Furniture and Devices

For each of the three rooms examined in detail, a dedicated sheet has been developed outlining the environmental requirements—namely, the optimal parameters for indoor comfort (such as appropriate temperature ranges, relative humidity, air exchange rates, lighting levels, noise thresholds, etc.), as well as dimensional guidelines (e.g., minimum widths of passageways and maneuvering areas, optimal heights for the placement of assistive devices, etc.) that influence the ergonomic quality of the space.

This set of information (reference values) represents the expected quality targets for spaces intended for home care. It defines the scope of the design response and guides the subsequent selection of technical elements, furniture, and devices, which—through their performance—must ensure the optimal conditions for carrying out care activities by both care recipients and caregivers (Fig. 4).

Each environmental unit is accompanied by a graphic representation aimed at providing an overview of the relevant environmental requirements, while leaving room for variability in possible alternative configurations (Fig. 5). This representation serves to highlight the technical elements,[5] components, furnishings, and devices that may be affected by future transformation interventions. Within the guidelines, these elements represent the starting point for a more detailed analysis, which is developed through specific in-depth technical sheets as described in the following section.

---

[5] Such as walls, perimeter and internal partitions, suspended ceilings, interior and exterior fixtures, flooring, wall coverings, lighting devices, and other system terminals.

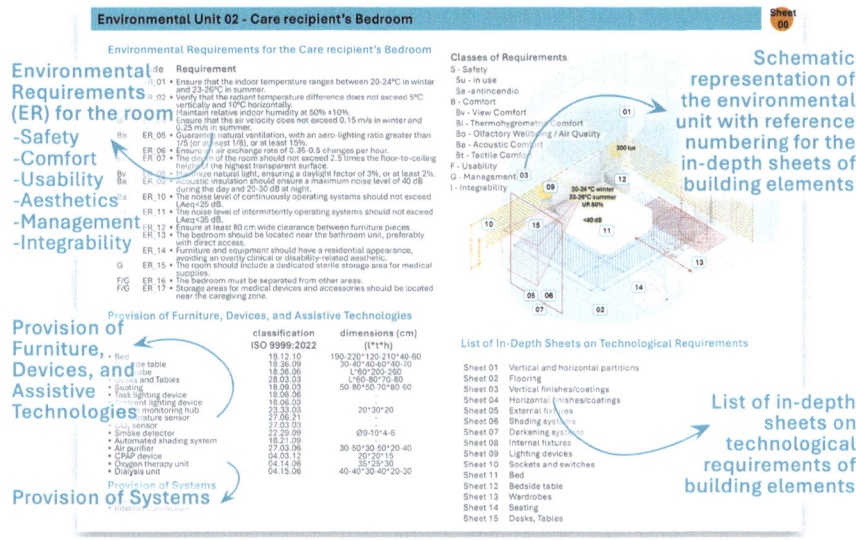

Fig. 4 Environmental requirements sheet for the Environmental Unit—Care recipient's bedroom

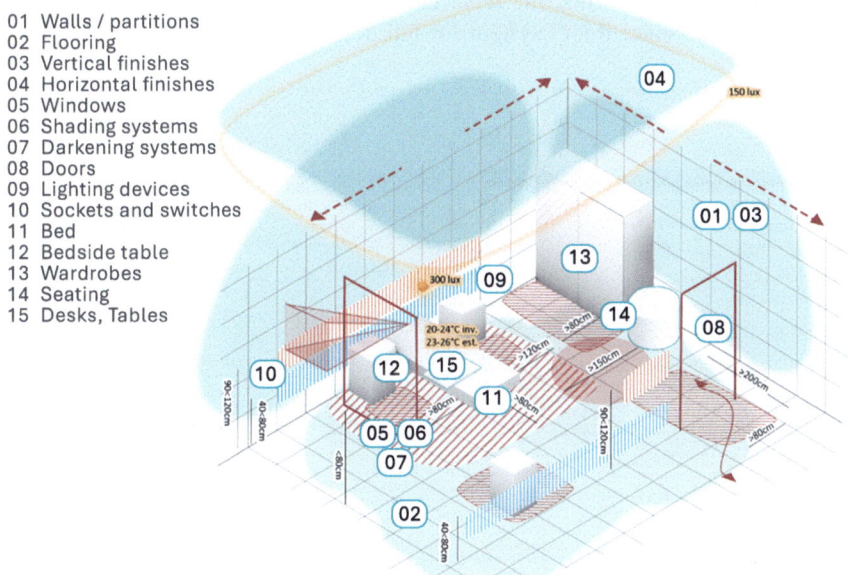

01 Walls / partitions
02 Flooring
03 Vertical finishes
04 Horizontal finishes
05 Windows
06 Shading systems
07 Darkening systems
08 Doors
09 Lighting devices
10 Sockets and switches
11 Bed
12 Bedside table
13 Wardrobes
14 Seating
15 Desks, Tables

Fig. 5 Graphic representation of the environmental unit "E.U. Care recipient's bedroom"

4.4 Technological Requirements of Technical Elements and In-Depth Technical Sheets

For each identified technical element, component, furnishing, and device, a dedicated in-depth technical sheet has been developed to define its specific performance-based requirements.

This set of information can be used to develop a range of alternative intervention scenarios, aimed at evaluating different strategies for adapting spaces to home care. Through the use of these in-depth sheets, it is possible to preliminarily identify key attributes related to technical elements, finishes, and the integrability of furniture and devices, as well as to anticipate the impact of the planned interventions.

An additional recommendation provided by the guidelines concerns the formulation of alternatives that allow for the concurrent comparison of different options. This supports decision-making based on specific needs and priorities related to the type and scale of intervention.

As an example, in the technical sheet dedicated to flooring within the care recipient's bedroom, in addition to defining the technological requirements that support the choice of appropriate materials, three possible intervention scenarios are proposed: the overlay of a new flooring system on the existing surface; the complete replacement of the existing flooring; the replacement of the flooring combined with the installation of a "smart floor" system for fall detection in the home environment (Fig. 6).

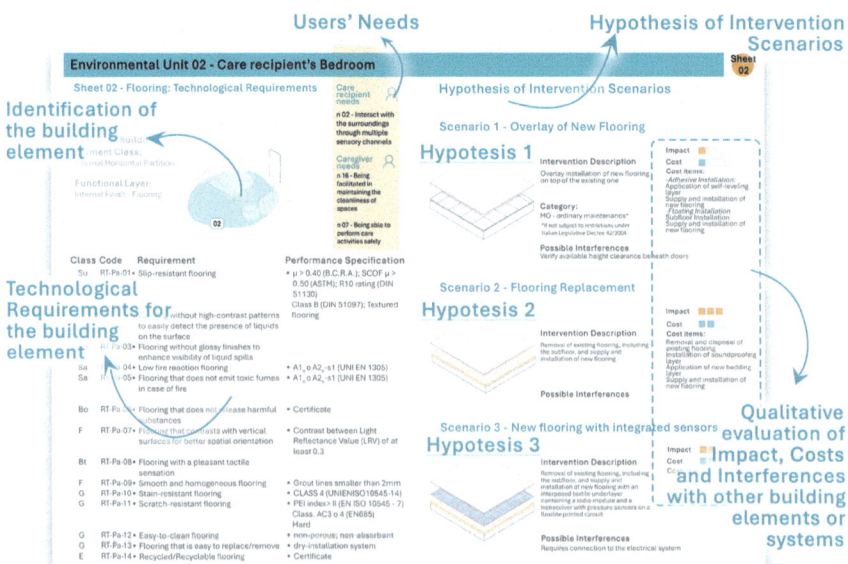

Fig. 6 In-depth technical sheet: technological requirements for the flooring of the environmental unit "E.U. Care recipient's bedroom", and proposed adaptation intervention scenarios

For each proposed alternative, the intervention is described in terms of its nature, referencing its classification under Italian building regulations (e.g., routine maintenance, extraordinary maintenance, etc.), as well as any applicable constraints.

Furthermore, each alternative is qualitatively assessed based on the following criteria:

- impact of the transformation, depending on whether it affects furniture, finishes, structural components, the envelope, or systems;
- costs, listing the main cost items involved in the intervention and specifying whether they fall under eligible categories for financial incentives currently available in Italy;
- interference with other technical elements and/or systems.

The proposed design solutions related to each intervention scenario are compiled in a comparative matrix, made available as a spreadsheet tool that can be queried by environmental unit, requirement, or by the specific technical element, functional layer, furniture, or device under consideration. This structure is intended to facilitate the selection of the most suitable options based on specific design conditions.

5 Conclusions

The growing attention to home care, promoted by recent healthcare reforms in response to demographic changes and supported by the spread of assistive technologies, calls for a critical rethinking of the role of the domestic space. The home can no longer be regarded solely as a place of everyday life, but must also be reimagined as a space for care and as a strategic node within the broader territorial network of social and healthcare services—one in which care activities can be carried out in conditions of safety, comfort, and respect for the needs of both older care recipients and their caregivers.

In this context, the design of domestic spaces is required to meet a complex system of needs that includes not only physical accessibility but also perceptual usability, technological integration, and the sustainability of interventions.

Through a demand-performance approach grounded in the person-environment relationship, the research reconstructed the framework of needs for both care recipients and caregivers. This served as the basis for identifying the main critical issues in existing domestic spaces and for defining a set of environmental and technological requirements applicable to the areas most involved in home care. These requirements formed the foundation for structuring the proposed guidelines and design strategies for home adaptation, which are intended to be transferable across different contexts and adaptable to varying levels of impact—from simple functional adjustments to more extensive renovation works.

The proposed strategies aim to go beyond the mere application of prescriptive measures, instead promoting the development of long-term design approaches capable of anticipating future needs and fostering inclusive, flexible living environments that can be easily personalized when specific care needs arise.

Based on the findings achieved thus far and presented in this contribution, future developments will focus on the implementation of digital tools to support designers, care providers, and policy makers in assessing the most effective solutions for reconfiguring existing housing. These tools are intended to overcome a fragmented view of domestic adaptation, integrating different areas of expertise within a collaborative ecosystem, and to help spread a new culture of housing adaptability that supports ageing-in-place and the transition toward "care-ready" environments.

Competing Interests This study was developed within the project funded by Next Generation EU—"*Age-It*—Ageing well in an ageing society" project (PE0000015), National Recovery and Resilience Plan (NRRP)—PE8—Mission 4, C2, Intervention 1.3".

The authors have no conflicts of interest to declare that are relevant to the content of this chapter.

References

1. National Research Council. (2011). *Health care comes home: The human factors*. The National Academies Press.
2. World Health Organization. (2001). *International classification of functioning, disability and health (ICF [ICIDH-2])*. Geneva.
3. Ministero della Salute. (2025). *Annuario Statistico del Servizio Sanitario nazionale. Assetto organizzativo, attività e fattori produttivi del SSN. Anno 2023*. https://www.salute.gov.it/new/sites/default/files/imported/C_17_pubblicazioni_3523_allegato.pdf
4. Vetrano, D. L. (Ed.) (2018). *L'assistenza domiciliare in Italia: chi la fa, come si fa e buone pratiche*. Italia Longeva.
5. Istat. (2021). *Le Condizioni Di Salute Della Popolazione Anziana in Italia. Anno 2019. Migliora La Salute Degli Anziani Ma Cresce La Domanda Di Cura e Assistenza*. https://www.istat.it/wp-content/uploads/2021/07/Report-anziani-2019.pdf
6. Kim, H., Ahn, Y. H., Steinho, A., & Lee, K. H. (2014). Home modification by older adults and their informal caregivers. *Archives of Gerontology and Geriatrics, 59*, 648–656. https://doi.org/10.1016/j.archger.2014.07.012
7. Ponikowski, P., Voors, A. A., Anker, S. D., Bueno, H., Cleland, J. G., Coats, A. J., Falk, V., González-Juanatey, J. R., Harjola, V. P., Jankowska, E. A., & Jessup, M. (2016). 2016 ESC Guidelines for the diagnosis and treatment of acute and chronic heart failure: The Task Force for the diagnosis and treatment of acute and chronic heart failure of the European Society of Cardiology (ESC). *European Journal of Heart Failure, 18*(8). https://doi.org/10.1093/eurheartj/ehw128
8. OECD. (2020). *Who cares? Attracting and retaining care workers for the elderly*. OECD Publishing. https://doi.org/10.1787/92c0ef68-en
9. Wu, S., Fu, Y., & Yang, Z. (2022). Housing condition, health status, and age-friendly housing modification in Europe: The last resort? *Building and Environment, 215*, 108956. https://doi.org/10.1016/j.buildenv.2022.108956
10. Márquez, G., & Taramasco, C. (2023). Barriers and facilitators of ambient assisted living systems: A systematic literature review. *International Journal of Environmental Research and Public Health, 20*(6), 5020. https://doi.org/10.3390/ijerph20065020

11. Carnemolla, P., & Bridge, C. (2019). Housing design and community care: How home modifications reduce care needs of older people and people with disability. *International Journal of Environmental Research and Public Health, 16*(11), 1951. https://doi.org/10.3390/ijerph16111951
12. Center for Aging Better. (2018). *Adapting for ageing: Good practice and innovation in home adaptations.* https://ageing-better.org.uk/sites/default/files/2018-10/Adapting-for-ageing-report.pdf
13. Newton, R., Adams, S., Keady, J., & Tsekleves, E. (2023). Exploring the contribution of housing adaptations in supporting everyday life for people with dementia: A scoping review. *Ageing and Society, 43*(8), 1833–1859. https://doi.org/10.1017/S0144686X21001367
14. Sanford, J. A., Pynoos, J., Gregory, A., & Browne, A. (2002). Development of a comprehensive assessment to enhance delivery of home modifications. *Journal of PT and OT in Geriatrics, 20*(2), 43–56. https://doi.org/10.1080/J148v20n02_03
15. Brand, S. (1994). *How buildings learn: What happens after they're built.* Viking.
16. Heidrich, O., Kamara, J., Maltese, S., Re Cecconi, F., & Dejaco, M. C. (2017). A critical review of the developments in building adaptability. *International Journal of Building Pathology and Adaptation, 35*(4), 284–303. https://doi.org/10.1108/IJBPA-03-2017-0018
17. Ulrich, R. (1984). View through a window may influence recovery from surgery. *Science, 224*(4647), 420–421. https://doi.org/10.1126/science.6143402
18. Kaplan, S. (1995). The restorative effects of nature: Toward an integrative framework. *Journal of Environmental Psychology, 15*(4), 169–182. https://doi.org/10.1016/0272-4944(95)90001-2
19. Lansley, P., Flanagan, S., Goodacre, K., Turner-Smith, A., & Cowan, D. (2005). Assessing the adaptability of the existing homes of older people. *Building and Environment, 40*(7), 949–963. https://doi.org/10.1016/j.buildenv.2004.09.011
20. Maryam, G., Murphy, C., Valenta, L., Bertram B., & Maxwell, D. (2021). Adaptable housing for people with disability in Australia: A scoping study. Australian Human Rights Commission. https://humanrights.gov.au/sites/default/files/document/publication/monash_-_adaptable_housing_2021_-_digital.pdf
21. Rabeneck, A. (2021). Housing adaptability: Some past lessons [Commentary]. *Buildings & Cities.* https://www.buildingsandcities.org/insights/commentaries/housing-adaptability-lessons.html

Open Access This chapter is licensed under the terms of the Creative Commons Attribution 4.0 International License (http://creativecommons.org/licenses/by/4.0/), which permits use, sharing, adaptation, distribution and reproduction in any medium or format, as long as you give appropriate credit to the original author(s) and the source, provide a link to the Creative Commons license and indicate if changes were made.

The images or other third party material in this chapter are included in the chapter's Creative Commons license, unless indicated otherwise in a credit line to the material. If material is not included in the chapter's Creative Commons license and your intended use is not permitted by statutory regulation or exceeds the permitted use, you will need to obtain permission directly from the copyright holder.

Social Day Care Centre: A New Architectural Model to Improve Elderly's Quality of Life

Maria Argenti, Fabio Cutroni, Domizia Mandolesi, Anna Bruna Menghini, Maura Percoco, and Francesca Sarno

Abstract The ageing population in Italy calls for a reassessment of housing and care models for the elderly, particularly through the development of Social Day Care Centres (SDCCs). These centres are envisioned as key elements of urban life, addressing the complex needs of the elderly by combining social interaction, healthcare, and community integration. Central to the SDCC model is the idea of 'ageing in place', which aims to reduce social isolation and promote well-being by enabling the elderly to remain in their homes and communities. The SDCCs are designed for a wide spectrum of elderly users, from those who are independent to those with varying degrees of reduced autonomy or cognitive impairments. These centres offer both social activities and healthcare services tailored to meet diverse needs. The design approach is inspired by European examples, especially from Spain, focusing on accessibility, flexibility, and seamless integration into the urban environment. Situated in central and peripheral urban areas, SDCCs encourage intergenerational interactions, foster social cohesion, and incorporate digital health monitoring tools. They serve as a critical component of community-based healthcare systems. The adaptable

---

M. Argenti · F. Cutroni · A. B. Menghini · M. Percoco · F. Sarno (✉)
Department of Civil, Building and Environmental Engineering, Sapienza University of Rome, Rome, Italy
e-mail: francesca.sarno@uniroma1.it

M. Argenti
e-mail: maria.argenti@uniroma1.it

F. Cutroni
e-mail: fabio.cutroni@uniroma1.it

A. B. Menghini
e-mail: annabruna.menghini@uniroma1.it

M. Percoco
e-mail: maura.percoco@uniroma1.it

D. Mandolesi
Department of Planning, Design, Architectural Technology, Sapienza University of Rome, Rome, Italy
e-mail: domizia.mandolesi@uniroma1.it

© The Author(s) 2025
T. Ferrante and M. Sacco (eds.), *Habitable Future*,
SpringerBriefs in Applied Sciences and Technology,
https://doi.org/10.1007/978-3-031-95735-2_13

nature of these centres allows them to be integrated into different urban contexts, ranging from standalone buildings to extensions within existing facilities. This flexibility ensures that SDCCs can evolve alongside societal changes, contributing to an inclusive and sustainable framework for elderly care in urban settings.

Keywords Social Day Care Centre · Architecture for community and health · Public health policy · Intergenerational perspective · Age-It

1 Living the Third Age Together

1.1 A New Perspective on Day Centres for the Elderly People

Italy's population is ageing. Data, statistical forecasts, everyday observations, and personal experiences, all confirm this reality: childless couples, families—including single-parent households—with only one child, elderly people living alone, and "young" seniors in good health. This evolving social composition is increasingly evident in urban and rural contexts, with slight variations between Northern, Central, and Southern Italy.

The analyses, estimates, and envisaged scenarios [1] inevitably challenge architectural design and urban planning, requiring immediate responses and proactive envisioning of future realities. It is essential to rethink the models of living and housing for the elderly in Italy. This need arises from recognising that ageing should not be seen as an exceptional condition, but rather as a normal phase of life—one that today is an increasingly significant part of an individual's life cycle.

The following pages are the partial result of a research project[1] carried out to address the need to integrate architectural quality, social engagement, and healthcare provision. This goal is pursued through the definition of guidelines for designing innovative Day Care Centres, specifically Social Day Care Centres. These centres are envisioned as new urban focal points, representing a hybridisation of existing models, on one side Social Protection Day Centres, on the other side Senior Centres.

Currently, Social Protection Day Centres assist individuals facing various challenges, distinguishing them from Day Centres for non-self-sufficient elderly people or those with dementia. Their primary aim is to prevent social isolation and loneliness while supporting people with reduced autonomy who live at home. These centres

---

[1] The survey is carried out by Task 1.4 of WP1 of Spoke 9 of the Age-It Extended Partnership. It is linked to Mission 6 (Health) of the National Recovery and Resilience Plan, which also promotes the social component within the "health system." From this perspective, the study aligns with the M6C1 component of the same Mission (Proximity networks, facilities, and telemedicine for territorial healthcare), which aims, among other objectives, to modernise health facilities, making them more digital and inclusive. It also seeks to strengthen preventive measures and local services while promoting research.

serve as an essential resource for those dealing with daily life, offering care and social services and, in the end, alleviating the burden on families and caregivers.

The latter, on the other hand, are focused on promoting socialisation for those people who still live independently and aim to keep the elderly active, encouraging their participation in social life.

Taking these considerations into account, and despite the usefulness of the existing centres, there is a need to define a new hybrid model, the Social Day Care Centre: a social assistance and recreational service for the elderly, intended for different types of users, with different levels of autonomy, aimed at supporting the elderly population in staying in their life environment, according to a model focused on the concept of 'ageing in place'. This approach seeks to prevent the elderly from leaving their home, thus avoiding social marginalisation, disconnection from the community, and the imposition of a life in residential care facilities. The goal of these new Day Centres is to support the elderly in the ageing process, ensuring they experience it as much as possible in a state of psychophysical well-being. It also aims to encourage prevention and health promotion, indirectly supporting primary healthcare services.

Therefore, these centres should provide opportunities for social interaction to foster interpersonal relationships, particularly for individuals who are alone and at risk of social exclusion, while also promoting an active and socially useful lifestyle. At the same time, they will offer rehabilitation programmes and activities to maintain both motor and cognitive skills.

To better define the characteristics that these new centres must fulfil, it is essential to reflect on the profile of the 'future elderly', individuals currently in their 50s and 60s. This generation is generally in good health, leading active lives, with more complex needs than the previous generation. They care about both their physical and cultural well-being and are familiar with digital information and communication technologies. However, they are also more likely to experience loneliness due to the increasing predominance of small families and more precarious economic conditions, alongside increasingly restricted levels of public social and healthcare spending.

In addition, especially in medium-to-large cities, there is a tendency towards a reduction—or even the lack of—constant neighbourly relations, which were always present in the past.

Cities must take responsibility for these social and demographic changes; they must be rethought to promote an inclusive system able to respond to both current and future needs.

For these reasons, the Social Day Care Centres are intended as part of a widespread and integrated system within the urban fabric, a 'proximity' service encompassing both day care and social interaction functions, not only for the elderly but also for individuals of different ages, as well as from diverse socio-economic and cultural backgrounds.

1.2 Semi-residential Care in Italy

Day care centres for the elderly are a type of public service that is not clearly codified and regulated.

They originated in Italy in the 1980s[2] in the wake of the international debate and were mainly intended for autonomous elderly people or those with initial limitations of autonomy. They originally provided mainly socialization activities, but when the target population gradually began to change in later years, the Centres mainly addressed non-self-sufficient elderly and/or those with dementia.

The D.P.C.M. of November 29, 2001 [2], by which the LEAs (Essential Levels of Care) were introduced, clarified that semi-residential care is to be understood as a basic level that every territory must guarantee for the non-self-sufficient elderly. The D.P.C.M. of January 12, 2017 [3], confirmed the previous norm, reaffirming that semi-residential care guaranteed by the National Health Service is for non-self-sufficient people with low need for health protection. Three types of Day Care Centres are provided in this regulatory framework: those for non-self-sufficient elderly, those for patients with dementia, and those of social protection for elderly in need of socialization and support.

While the first two types are recognised as social-health facilities to be guaranteed throughout the national territory, the implementation of the third one, with a social character, is left to the political discretion of the different territories.

For the latter type—Social Protection Day Care Centres—it is not possible to refer to a national model of organization, since it can take on configurations adaptable to the needs of the recipients and the contexts to which they belong [4]. In fact, for these centres, the minimum allocation of space or the optimal number of guests is regulated by legal provisions that differ from Region to Region.

While the absence of rigid regulations creates an inevitable territorial disparity, this elasticity allows for autonomy in a programmatic setting that can respond more nimbly to the ongoing demographic and socio-cultural changes. Moreover, it allows a certain freedom in planning choices, which is especially necessary in a logic of reuse of the existing building stock. In developing design guidelines, regulatory flexibility has allowed to also refer to case studies outside of Italy. The design experiences taken into consideration therefore belong to the broadest contemporary European panorama, where examples are increasingly numerous and characterised by definitely innovative architectural approaches.

---

[2] The legislative measures enacted in Italy since Law No. 67/1988, are aimed at regulating, at the national level, the management and organization of the services of facilities intended for the elderly population, with varying degrees of self-sufficiency, and to encourage the maintenance of full autonomy to lead an affectively and socially active existence.

1.3 Good Design Practices

In Europe, there are environmental and cultural contexts which are similar to the one we have in Italy—for example the Iberian Peninsula—where social policies and design research on living spaces for the elderly are highly advanced [5].

Some *Centros de Día para Mayores* in Spain stand for paradigmatic examples, due to the cultural and social proximity between Spain and Italy, and also because in some Spanish regions the average age of the population and the percentages at risk of poverty and social exclusion are comparable to those in our country.

An in-depth study of these examples allows to highlight significant design features and aspects as well as recurring themes and peculiarities for each case.

Centros de Día are generally limited in size (to preserve a human and domestic scale) and are typically single-storey (to facilitate accessibility). Although small, they play a pivotal urban, social, and symbolic role. Located in the heart of the city or in its outskirts, they emphasise the public nature of their function. Some of them, with their prominent features, stand as landmarks in the landscape; others, thanks to their articulation, create outdoor spaces open to a public use; others, lastly, interact even more with their surroundings, like freely crossable 'urban gates.'

In these buildings, the relationship between indoor and outdoor spaces is extremely important, achieved mainly through visual and functional continuity. Similarly, the choice of construction and finishing materials is highly considered, favouring natural materials, such as wood and stone, as well as colours, to give a sense of domesticity, vitality, and harmony to the rooms.

These considerations lead to the guidelines for the design of new Social Day Care Centres, capable of positively addressing the needs of socialisation, autonomy, and well-being of the users.

2 Guidelines for the Design of Social Day Care Centres

2.1 Plurality of Users and Community Cohesion

Although maintaining a healthy and active life is a central point from a health and socio-cultural perspective, architecture and urban planning can still make an important contribution since life is lived in 'places.'

Based on the existing publications, on the requirements set forth by national and regional regulations, and on some paradigmatic examples, it is possible to infer guidelines for designing innovative day centres. These architectural guidelines are mainly referred to new constructions; however, they are also suitable for the adaptive reuse of existing, disused or underused buildings.

As already pointed out, the current legislation makes a clear difference between day centres, separating those for self-sufficient users from those for non-self-sufficient users or with cognitive impairments.

Considering the data released by the OECD, with member countries probably having 32 million people with dementia by 2040, it is crucial to promote a policy for the prevention of cognitive diseases and to encourage integration processes for those elderly already affected by them.

Therefore, the mix of users, although it involves more significant management challenges, represents a positive factor for the elderly, especially for those with cognitive impairments, who are more prone to isolation. To ensure an adequate coexistence of users, the architectural requirements have also to consider the basic needs of those centres intended exclusively for patients with Alzheimer's or other cognitive disorders [6].

First and foremost, spaces must be flexible both to fit the changes that may come out with aging and to allow individuals with varying levels of environmental and social interaction capabilities to coexist without interference.

In addition to a mix of users related to their physical condition, Social Day Care Centres should also accommodate a variety of users from a socio-economic and cultural standpoint, fostering social cohesion.

This can be achieved firstly by offering diverse recreational activities involving people of different ages and socioeconomic backgrounds and secondly by scattering these centres throughout the neighbourhoods, thus helping to maintain community cohesion over the years. In particular, the presence of young people and children aims to foster a relationship among different generations whose gap, according to recent data, is expected to widen increasingly.

Overall, Day Centres for users with varying degrees of autonomy are optimal when arranged as integrated systems within the urban context, providing spaces for socialising—including intergenerational interactions—and cultural and occupational activities, together with care and health monitoring functions to ensure autonomy and psycho-physical well-being, prevent the onset of diseases, and help mitigate the burden on the national healthcare system [7].

Regarding the number of users, a simultaneous presence of 30–40 people is optimal, in order to create a small community and facilitate the activities in small groups.

2.2 Framework of Requirements to Drive the Project

Analysing the framework of requirements related to the different users the innovative Social Day Care Centre is addressed to, the location criteria, settlement principles, and architectural and functional features, supposed to drive the design, are derived.

Below the list of some users' needs considered as priorities and the related design requirements.

Belonging	• Sense of belonging to the community
	Social Day Care Centres should be in a way that:
	  – Preserves social, cultural, and historical ties with a place, context or community
	  – Safeguards the connection of the elderly to their family environment
	  – Respects customs and habits on which the emotional well-being of feeling part of the place where you live is based on
	• Sense of belonging to the space
	Social Day Care Centres have to provide environments that:
	  – Ensure a family-like atmosphere
	  – Promote inclusion and integration for all types of users
	  – Offer points of reference which create a sense of safety and comfort in using the spaces

Sociality	• Relational exchanges within the user community
	Social Day Care Centres should provide spaces that allow different users to interact, based on the following characteristics:
	  – Socio-demographic (age, gender, marital status)
	  – Sociocultural (kind of education, socioeconomic status)
	  – Physical-psychological (health conditions)
	  – Lifestyles (hobbies, personal interests, life experiences)
	• Relational exchanges between user communities and socio-demographic context
	To foster a proactive exchange between elderly users and people of other different ages (adults, young people, children) coming from outside, it is crucial to provide functional areas which may welcome:
	  – Daily activities aimed at developing educational and volunteering projects
	  – Mutual support activities that encourage the exchange of skills and knowledge

Cultural enrichment	• Support to personal culture It is essential to provide dedicated spaces where individuals can engage in leisure and self-enrichment activities, such as reading or cognitive support activities, including memory exercises and games • Involvement in collective cultural, artistic, and creative initiatives To foster an inclusive and culturally stimulating environment where individuals can take on 'actor' or 'spectator' roles in various activities, the Social Day Care Centres should include wide yet flexible meeting areas suitable for accommodating more or less large groups of people

Well-being	To support the most active life possible, the Social Day Care Centres must promote physical and psychological well-being
	• Physical well-being
	  Spaces for physical activities, such as rehabilitation gyms, and spaces to support physical health, allowing for example occupational therapy, should be provided
	• Psychological well-being
	  It is essential to offer spaces for stimulating activities, such as intellectual or manual workshops, while incorporating—where possible—outdoor areas, given the positive impact of a natural environment and greenery on the individual's psychological well-being, regardless of age

(continued)

(continued)

Care	• Prevention Spaces where to evaluate individual lifestyles of the elderly and, if necessary, modify them by identifying habits that may positively or negatively have impact on physical and mental health should be provided. These spaces can be used also by groups of elderly people for seminars or workshops aimed at educating and raising awareness of healthy behaviours, including, for example, proper hydration and nutrition • Monitoring of lifestyles and physiological parameters It is recommended to provide spaces that may allow: – The regular measurement of physiological parameters using instruments such as glucometers and sphygmomanometers – Support in integrating easy-to-use equipment into the daily lives of elderly individuals to ensure continuous monitoring (e.g., smartwatches, wearable sensors)	
Autonomy	In order to improve quality of life and reduce dependence on external assistance, the Social Day Care Centres should provide areas for activities which may support the maintenance of personal autonomy while allowing as much freedom of movement as possible • Self-sufficiency in personal care To promote autonomy of elderly individuals, facilities should include spaces suitable for: – Demonstrating simplified techniques to manage daily tasks – Teaching personalised routines to facilitate personal care activities • Independence in the Use of Spaces To ensure ease of use, the facilities should: – Be free from physical and perceptual obstacles – Have a clear and logical spatial layout – Allow easy identification of the different functional areas and of the objects or furniture within them – Feature appropriate signage	
Welcoming spaces and privacy	• Spatial versatility and space perception As the Social Day Care Centres are designed to accommodate elderly people with different physical and cognitive abilities, as well as individuals from other age groups, it is essential to implement design strategies that cater to different needs. These strategies include: – Defining well-arranged, easily identifiable, and reassuring spaces – Creating a domestic-like atmosphere and scale – Allowing for the personalisation of spaces – Using colours, signage, and familiar objects to aid orientation Even materials can play a key role in fostering an empathetic, familiar, and comfortable environment, thus giving a strong identity to the space. This can be achieved through: – Natural materials that convey authenticity and perceptive warmth – Tactile and soft textures to promote feelings of comfort – Construction details that enhance the domestic character of the space • Privacy To guarantee dignity and privacy of individuals while fostering comfort and trust towards the staff, it is crucial to provide suitable spaces for small groups or individuals, particularly for those activities related to prevention and health monitoring	

(continued)

(continued)

Safety and accessibility	• Safe environments
	  Safety of spaces means a higher feeling of freedom, reducing the need for constant staff supervision or restrictions on individual use. Security devices should be non-invasive and, as much as possible, discreetly integrated in order to maintain a 'domestic' atmosphere rather than a highly controlled technological environment
	• Autonomous Accessibility
	  To meet the needs of easy accessibility and adequate safety, the Social Day Care Centres should preferably be on the ground floor and on a single level, both in the case of new construction and adaptive reuse. The layout should avoid as much as possible differences in height, ensuring flat floors in both indoor and outdoor spaces. In the case of adaptive reuse, short, slightly sloped ramps should ensure accessibility. Additionally, spaces should be immediately visible through proper natural and artificial lighting and free from danger or physical and psychological barriers inside and outside

(continued)

(continued)

Environmental comfort	• Microclimate In Social Day Care Centres natural and technological solutions should be combined to properly manage climatic conditions (temperature, humidity, ventilation, and solar radiation), and achieve the best thermo-hygrometric comfort. These solutions should also be able to reduce the environmental impact of the building and, therefore, focus on energy efficiency Among the possible solutions we may consider: – Natural shading (e.g., planting of trees) to regulate both outdoor and indoor temperatures – Optimisation of solar radiation (building orientation, reduction of summer overheating using mobile or adjustable shading) – Provision of opposing openings to foster natural cross ventilation and reduce the concentration of pollutants – Possible vegetation on the roofs – Use of light-coloured and reflective (not dazzling) materials to reduce heat accumulation – Insertion of natural barriers to mitigate cold winds – Use of technological systems to control air quality, humidity, and indoor temperature • Lighting Room lighting should ensure visual comfort, safety, and energy efficiency. Natural light sources, such as large windows with shading devices, should be preferred, thus achieving as uniform a brightness as possible to avoid shadows and dark areas Regarding artificial lighting, the following is recommended: – Warm-neutral light to create a welcoming and natural environment – Colour Rendering Index (CRI) between 80 and 89, the best for home environments, to ensure clear perception of colours and facial expressions – Diffused light for large rooms and common areas, avoiding sudden transitions between bright and shaded areas – Soft light to avoid disorientation – Direct and shielded light for reading areas or spaces for manual activities – Motion sensors in transit areas for energy saving – Glare-free lighting for healthcare spaces • Noise In addition to choosing suitable locations protected from external noise, it is necessary to adopt proper strategies to achieve the best control of both the sound level and the activities arrangement, thus making spaces more serene and accessible, in particular considering users with hearing impairments or those suffering from cognitive disorders. Regarding the different spaces, the following is recommended: – Sound-absorbing materials on walls and ceilings – Soft furniture – Soundproof paving, using materials such as linoleum or vinyl – Partitions and separations made by movable walls or open bookcases to reduce sound propagation Regarding the activities arrangement, the distribution layout should be based on the following two recommendations: – Separate noisy areas (dining room, spaces for group activities) from quieter areas (relaxation areas, reading areas) – Arrange spaces to avoid gatherings and confusion

Based on the above requirements, the design indications focus on location and functional aspects, which are fundamental to any architectural intervention.

The aim is to provide design principles that support the definition of this new architectural model—hybrid and versatile.

These indications are intended for structures situated in metropolitan contexts or established urban areas and are conceived to accommodate a mix of functions that can adapt to aggregation systems of varying complexity.

2.3 Design Indications

Location. Social Day Care Centres form the basic units of a system spread across the territory. The model underlying these guidelines is an architectural organism, part of a multipolar system, contributing to a widespread network of proximity [8]. It aims to integrate and strengthen the social and healthcare facilities established by national laws and those foreseen by the National Recovery and Resilience Plan to enhance community-based healthcare. In order to achieve such a widespread accessibility, it is essential to ensure a homogeneous distribution within the cities, considering the neighbourhood scale as a reference [9]. Therefore, these centres shouldn't be isolated but rather located within urbanised areas, either central or peripheral, well connected to the surrounding environment, thus ensuring a high level of social participation, together with the involvement of the local community [10, 11] (Fig. 1).

Fig. 1 Schematic representation of the city-social day care centre system

To integrate them as much as possible within the social context, it is desirable to choose available spaces or disused buildings near gathering places, such as:

- Parks, green areas, and squares
- Essential public services (schools, administrative offices)
- Cultural and recreational spaces (theatres, cinemas, and libraries)
- Assistance services (diagnostic centres, hospitals)
- Places of worship
- Markets and commercial areas

To make Social Day Care Centres easily accessible to elderly users and their caregivers, they should be close to:

- Public transport hubs (railway or underground stations, bus stops)
- Pedestrian or bicycle paths

Furthermore, to foster a meaningful relationship between users and nature—essential for ensuring a feeling of well-being and harmony—it is advisable to adopt strategies that facilitate interaction between the built and natural environments. These strategies include:

- Placing the centres within parks, green areas, or gardens
- Embedding shared green spaces (courtyards, patios) in the project
- Providing views towards gardens, natural landscapes, or open urban spaces
- Integrating natural elements within the building
- Arranging vegetable gardens, greenhouses, and community cultivation areas
- Ensuring proximity to spaces that promote biodiversity experiences.

Concerning the architectural scale, these Centres can be housed in standalone structures, within buildings that accommodate other community functions, or in areas of residential buildings with independent access. Specifically, positioning the Centre within a Nursing Home (RSA) [12] allows it to be open to the neighbourhood and facilitates the 'continuum of care,' even as the elderly person's health condition changes. However, this integration may make the Centre's role as a support for elderly individuals in their own homes, as an alternative to the RSA, less perceptible.

Space Planning and Functional Organization. In Social Day Care Centres, conceived as an innovative 'territorial public service' two essential functional macro-areas exist: 'social' and 'care.' The social area encompasses recreational and cultural activities open to a broad and diverse community context (Fig. 2). The care area, on the other hand, focuses on elderly care services. In this way, these Centres serve as reference points for home support within the continuum of care (Fig. 3).

Social Day Care Centre: A New Architectural Model to Improve ... 153

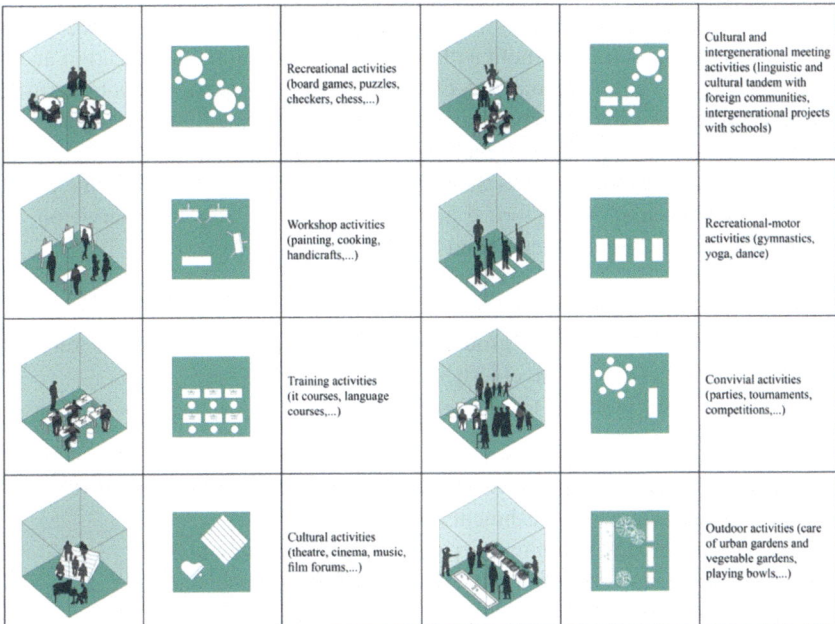

Fig. 2 The 'social' functions to be incorporated into future Social Day Care Centres

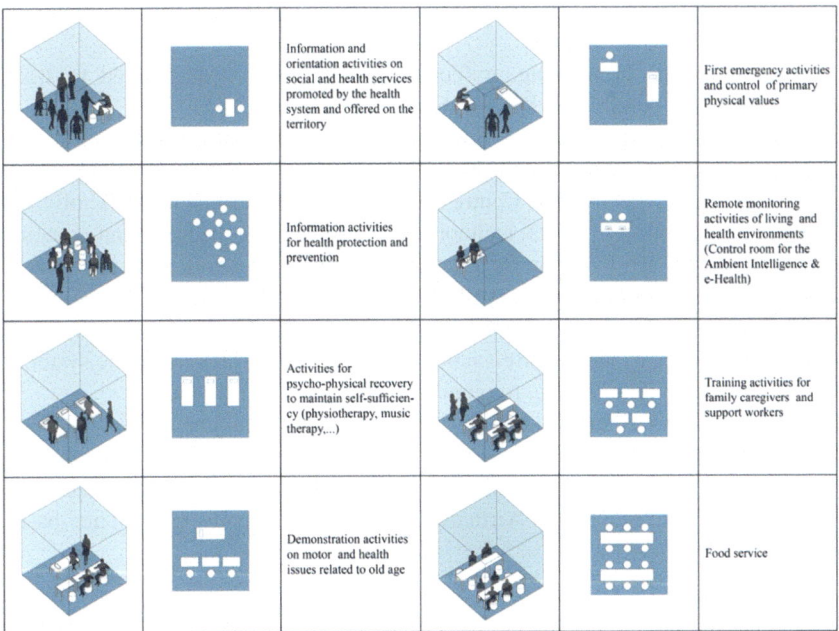

Fig. 3 The 'care' functions to be incorporated into future Social Day Care Centres

By combining these two aspects, Social Day Care Centres address the need for a public service that functions as both a place for intergenerational aggregation and a day-care centre, all within a framework of adaptive reuse of abandoned or underused real estate. Other roles may also be associated with the aforementioned macro-areas, extending beyond the Centre itself and operating at the territorial level. These can include voluntary activities such as the protection and management of public green spaces, the enhancement of historical and artistic heritage, school supervision, as well as food and clothing collection and distribution.

As a result, the Social Day Care Centre is conceived as an open system, promoting integration with the environmental and social context, encouraging exchange and comparison between different age groups, and providing spaces for intergenerational activities, primarily between the elderly and children.

The spatial aspects practical to conveying the message of openness, inclusiveness, and sharing are represented by:

- The size and quality of spaces for collective activities (meeting rooms for local associations, play areas, refreshment points)
- A direct connection with community gardens and vegetable plots
- Flexibility and versatility of interior spaces, and perceptual continuity

Smart Centre Functions. The multiple functional framework envisaged for the Social Day Care Centres also includes healthcare and protection functions, collectively referred to as 'care' (e.g., first aid, monitoring of basic physical parameters, and remote health monitoring), which can be carried out through digital therapeutic and health tools.

To facilitate this monitoring, a 'control room' can be designated to track individuals' health conditions (Digital Healthcare).

From a broader perspective, the Social Day Care Centre also serves as a neighbourhood hub, enabling remote monitoring of the living conditions of people living alone. From this perspective, the Centres become hubs in a virtual network, offering remote assistance through digital technologies, transforming the home into a secure space and an essential pillar of the healthcare infrastructure network. Among the technologies suitable for these purposes (Ambient Intelligence, AmI), we can consider those focused on security (such as smart sensors, household appliance control, and remote monitoring), as well as life-saving alarm devices and telemedicine systems, which can be used, for example, to monitor drug intake.

Therefore, thanks to current and future technology and equipment, the neighbourhood senior centre or the daycare centre for people with psycho-motor vulnerabilities becomes a new Smart Centre—an integrated hub within the urban fabric, capable of connecting, both physically and 'virtually,' a segment of society, without age-based constraints [13].

2.4 A Diffuse, Open, Composite, and Modular System

It is evident that the strategies for shaping the project cannot be one-size-fits-all but should vary according to the specific contexts and needs of the target population, leading to structures that are diversified in terms of typology, size, and design.

The functional areas may adopt variable configurations and proportions depending on factors such as the type and number of users, the size of the lot or existing building (in the case of reuse), and the presence or absence of social and healthcare facilities in the neighbourhood. Additionally, the intended user mix and potential variations in user characteristics and behaviours due to socio-demographic changes necessitate flexibility and the capacity to adapt spaces.

A single Social Day Care Centre can accommodate all priority functions or only a selection of them. It may be integrated with other social and healthcare services within its context or operate independently to address local shortcomings.

As such, it is conceived as a system composed of parts that can be both autonomous and interdependent, connected either physically or within the broader urban framework. This approach ensures adaptability to different contexts and flexibility within the system, ranging from a 'major system', which includes all functional areas, to a 'minor system', which provides only a few but still effectively supports the community (Figs. 4 and 5).

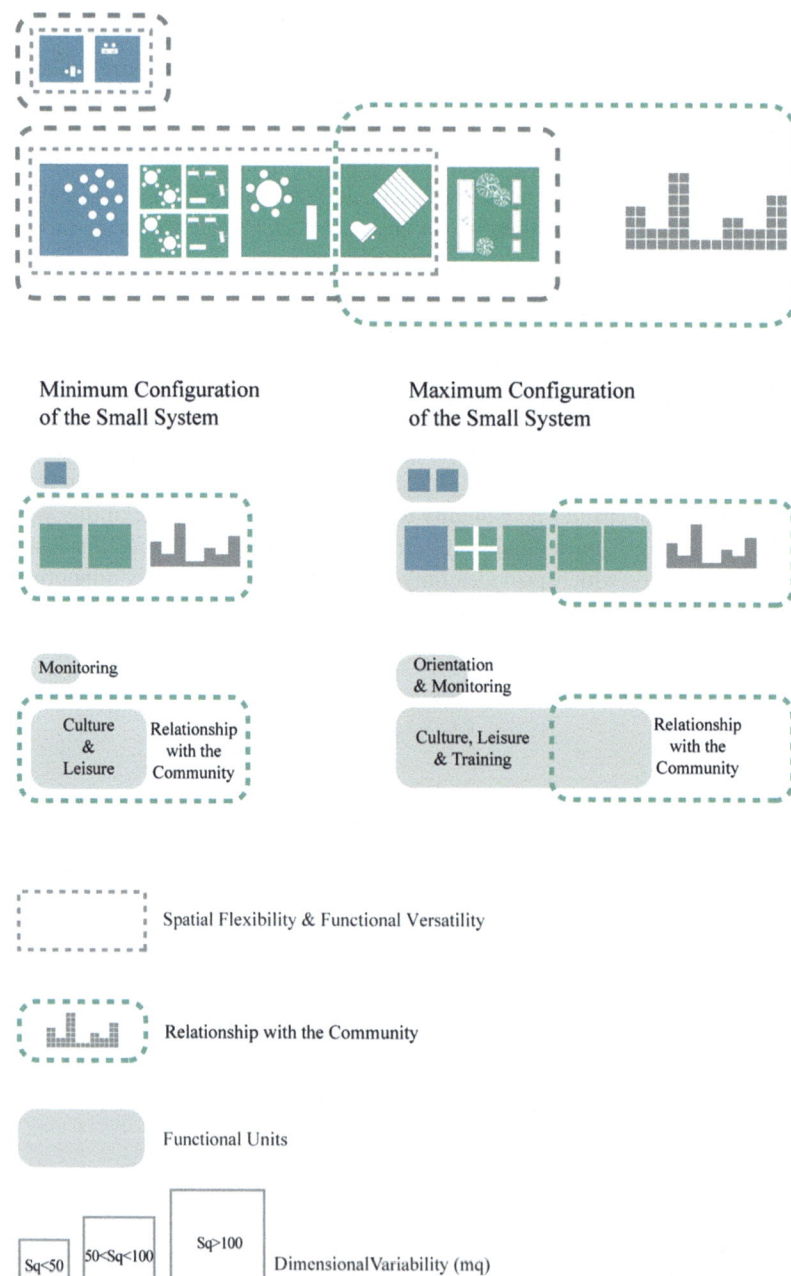

Fig. 4 Small model of Social Day Care Centre with variable complexity

Social Day Care Centre: A New Architectural Model to Improve ... 157

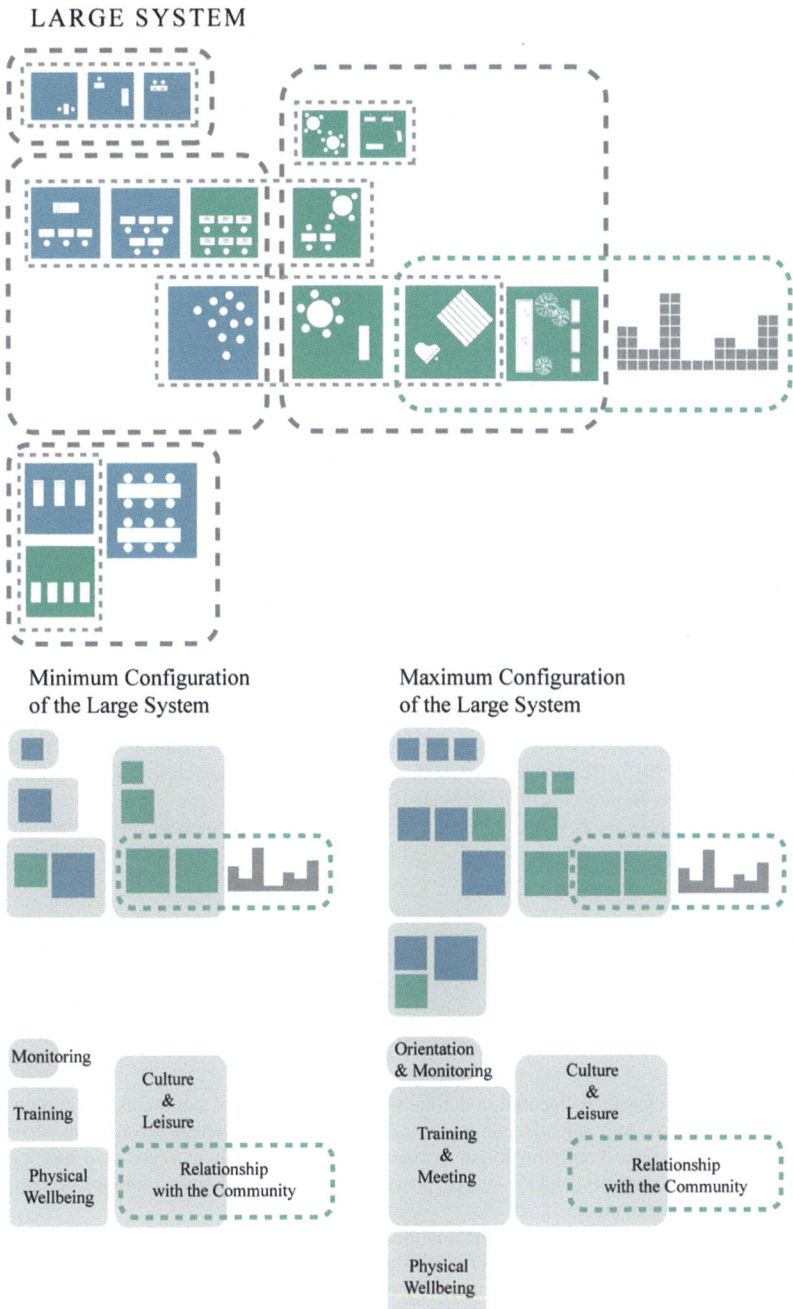

Fig. 5 Large model of Social Day Care Centre with variable complexity

Competing Interests This study was developed within the project funded by Next Generation EU—"*Age-It*—Ageing well in an ageing society" project (PE0000015), National Recovery and Resilience Plan (NRRP)—PE8—Mission 4, C2, Intervention 1.3".

The authors have no conflicts of interest to declare that are relevant to the content of this chapter.

References

1. ISTAT. (2022). *Previsioni della popolazione residente e delle famiglie* | base 1/1/2021. https://www.istat.it/it/archivio/274898
2. D.P.C.M. (2001). Definizione dei livelli essenziali di assistenza.
3. D.P.C.M. (2017). Definizione e aggiornamento dei livelli essenziali di assistenza, di cui all'articolo 1, comma 7, del decreto legislativo 30 dicembre 1992, n. 502. (17A02015).
4. Pesaresi, F. (2018). *Gli approfondimenti di NNA. Manuale del Centro Diurno*. Maggioli Editore.
5. Arup, HelpAge International, Intel, & Systematica. (2015). *Shaping ageing cities: 10 European case studies*. Arup.
6. VV.AA. (2023). Italian guidance on dementia day care centres: A position paper. *Aging Clinical and Experimental Research, 35*, 729–744.
7. VV.AA. (2017). Architecture and social innovation. *TECHNE: Journal of Technology for Architecture and Environment,* 14(7).
8. Buffel, T., & Phillipson, C. (2016). Can global cities be 'age-friendly cities'? Urban development and ageing populations. *Cities, 55*, 94–100.
9. Arup. (2019). *Cities alive: Designing for ageing communities*. Arup.
10. VV.AA. (2020). Public space. *TECHNE: Journal of Technology for Architecture and Environment, 19*(10).
11. Spadolini, M. B. (2013). *Design for better life: longevità, scenari e strategie*. Franco Angeli.
12. Mandolesi, D. (2015). Residenze per anziani. In Dall'Olio, L., & Mandolesi, D. (Eds.), *Manuale di progettazione. Residenze collettive*. Mancosu Editore.
13. Martinoni, M., & Sassi, E. (Eds.). (2013). *UrbAging. La città e gli anziani*. Tarmac Publishing.

Open Access This chapter is licensed under the terms of the Creative Commons Attribution 4.0 International License (http://creativecommons.org/licenses/by/4.0/), which permits use, sharing, adaptation, distribution and reproduction in any medium or format, as long as you give appropriate credit to the original author(s) and the source, provide a link to the Creative Commons license and indicate if changes were made.

The images or other third party material in this chapter are included in the chapter's Creative Commons license, unless indicated otherwise in a credit line to the material. If material is not included in the chapter's Creative Commons license and your intended use is not permitted by statutory regulation or exceeds the permitted use, you will need to obtain permission directly from the copyright holder.

The manufacturer's authorised representative in the EU is Springer Nature Customer Service Centre GmbH, Europaplatz 3, 69115 Heidelberg, Germany. If you have any concerns regarding our products, please contact ProductSafety@springernature.com

Printed and bound by CPI Group (UK) Ltd, Croydon, CR0 4YY

23/03/2026

02076360-0017